AWESOME ALGEBRA

METRO BOOKS
New York

An Imprint of Sterling Publishing
1166 Avenue of the Americas
New York, NY 10036

Conceived, designed, and produced by
Quid Publishing
Level 4 Sheridan House
114 Western Road
Hove BN3 1DD
England

Design by Lindsey Johns

www.quidpublishing.com

ISBN: 978-1-4351-6249-5

For information about custom editions, special sales, and premium and corporate purchases, please
contact Sterling Special Sales at 800-805-5489 or specialsales@sterlingpublishing.com.

Manufactured in China

2 4 6 8 10 9 7 5 3 1

www.sterlingpublishing.com

Image credits
- Corbis: 104, 128
- Creative Commons: 56, 59 © Anneke Bart, 80 © Stefano Bolognini, 94 (top) © Wolfgang Beyer,
  110 © Tieum512, 139 (inset) © Bob Lord
- Getty Images: 69 © DEA | M. Seemuller, 129 © Ullstein Bild
- Misc: 90 (bottom) © Luc Viatour (www.lucnix.be)
- Shutterstock: 3, 83, 86, 87 (all), 89, 113, 114, 127
- Wellcome Trust: 78 & 92 This image comes from Wellcome Images, a website operated by
  Wellcome Trust, a global charitable foundation based in the United Kingdom.

AWESOME
ALGEBRA

A TOTALLY NON-SCARY GUIDE TO ALGEBRA AND WHY IT COUNTS

Michael Willers

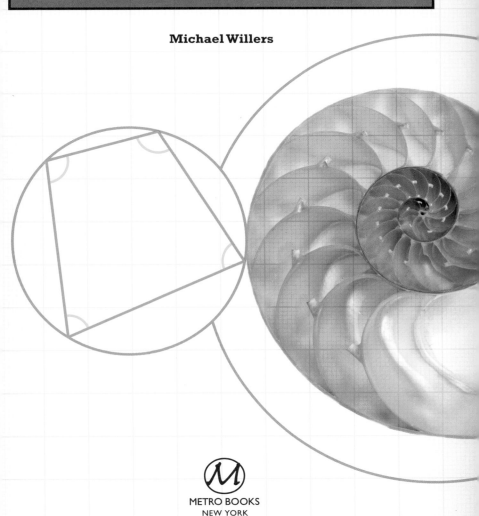

METRO BOOKS
NEW YORK

Contents

Introducing Algebra 6

1 THE BASICS OF ALGEBRA 12

What's in a Number? Part 1 14

What's in a Number? Part 2 16

Backstage with Pi 20

Playing by the Rules 22

Expressions, Equalities, and Inequalitites 24

The Power of Polynomials 26

Talking Trig 28

"The Laws of Nature are written in the language of mathematics."

—Galileo Galilei

2 CLASSICAL MATH 32

Profile: Pythagoras 34

Pythagorean Theory 36

Radical Mathematics 38

Profile: Plato 40

Platonic Solids 42

Profile: Euclid 44

Profile: Archimedes 46

Archimedean Solids 48

Profile: Eratosthenes 50

Profile: Diophantus 54

3 AN INTERNATIONAL LANGUAGE 56

Egyptian Mathematics 58

Completing the Square 60

Indian Mathematics 62

Profile: Brahmagupta 64

Math in the Middle East 68

Profile: Al-Khwarizmi 70

Profile: Omar Khayyam 72

Conics 74

The Quadratic Formula 76

4 THE ITALIAN CONNECTION 78

Profile: Fibonacci Part 1 — 80

Let's Get Graphical — 82

Profile: Fibonacci Part 2 — 84

The Golden Ratio — 88

Profile: Tartaglia and Cardano — 92

Imaginary Math — 94

Complex Arithmetic — 96

Conjugates — 100

Quadratics, Parabolas, and Complex Numbers — 102

Sierpinski's triangle: see Pascal's Triangles (pp. 116–21)

5 POST-RENAISSANCE EUROPE 104

Profile: René Descartes — 106

Graphing Lines — 108

Profile: Blaise Pascal — 110

Fun with Factorials! — 112

Permutations and Combinations — 114

Pascal's Triangle — 116

Pascal's Triangle Part 2 — 118

The Binomial Theorem — 122

Profile: Leonhard Euler — 124

Hit the Road — 126

Profile: Carl Friedrich Gauss — 128

6 TALKING IN CODE 130

Exponents Law — 132

Exponential Equations — 134

Logarithms — 136

Codes and Ciphers — 138

Caesar and Vigenère Ciphers — 140

Index — 142

Glossary — 144

Introducing Algebra

Mathematics means many things to many people. For some it represents all the beauty of the universe. Alfred North Whitehead (1861–1947), an English mathematician and philosopher, described pure mathematics as "the most original creation of the human spirit." For others, however, mathematics is a daunting subject, whether it takes the form of equations on a whiteboard or the fact that, however many times you try, you can never make your cash stretch far enough.

A Beautiful Kind of Magic

Of course, things aren't quite so black and white. On the one hand, even the biggest math-hater in the world can't deny the beauty of something as simple yet compelling as the Fibonacci sequence (see p. 85) when it occurs in nature; while, on the other, even the most avid mathmaniac would have to admit that many of mathematics' most beautiful ideas are shrouded in an air of mystery, making them inaccessible to most of the population.

This air of otherness can be intimidating. In fact, the fourth-century theologian St. Augustine of Hippo claimed that mathematicians had made "a covenant with the Devil to darken the spirit and confine man in the bonds of Hell." Maybe you might've agreed with him during a particularly difficult math class? However, it doesn't have to be like this. Yes, there are some difficult ideas, but there is also a lot that is beautiful. To study the nature of mathematics, we must consider what mathematics is really about, what makes it unique, and how it developed.

This book aims to help you on that journey. Along the way, we'll meet many interesting and challenging ideas, but our aim is to connect them to the everyday world; and where that can't be done, simply to sit back and marvel at their beauty.

Two of the five **Platonic solids**. These fascinating shapes have unique properties (see pp. 42–43).

What Is Mathematics?

Mathematics has been called the science of number and magnitude, the science of patterns and relationships, and the language of science. Galileo, the famous Italian scientist who lived from 1564 to 1642, claimed that "The Laws of Nature are written in the language of mathematics." In fact, mathematics is all these things. It's a growing, creative, and dynamic field of inquiry. The explosion of mathematics is often hidden from the general population, but that has changed in recent years. Scientific discovery and debate—for example, surrounding global warming—have led to a desire to understand the math that underlies it. The popular media have also played their part, and mathematics has been at the heart of such works as the Oscar-winning film *A Beautiful Mind* (an adaptation of the Pulitzer Prize-nominated book of the same name) and Dan Brown's bestseller *The Da Vinci Code*. Even psychedelic fractal posters reveal an appreciation of mathematical beauty, whether or not people choose to see the numbers behind the patterns.

In spite of this, some people still see mathematics as a static discipline, isolated from the real world. This is chiefly the fault of an education system that spends most of its time reviewing material developed millennia ago. That's not to say that these topics aren't important and interesting, but it's rarely conveyed just how fluid a subject mathematics is, let alone how it has developed and grown over time. Mathematics has a rich history and within these pages we'll meet many of the fascinating mathematicians who shaped it.

What's more, math continues to be shaped by remarkable individuals. Andrew Wiles, the mathematician who famously solved "Fermat's Last Theorem" in 1994, and the popularity of Simon Singh's ensuing book, are both proof that mathematics is still growing and changing.

The numbers of the **Fibonacci sequence** occur in many places in nature. In this case, there are twenty-one petals on the flower.

"The Laws of Nature are written in the language of mathematics."
—**Galileo Galilei**

A Short History of Mathematics

As the discovery of a 30,000-year-old wolf-bone notched with tallies of five attests, the history of mathematics is ancient indeed.

The revelation that mathematics is not the exclusive domain of the human race—for example, it has been shown that crows can distinguish between sets of up to four elements—demonstrates that the rudiments of counting occur in other creatures. What's more, it begs the question: which came first, humans or mathematics?

> **"**There is no royal road to geometry.**"**
>
> —Euclid

Given mathematics' long and colorful history, it's hardly surprising that the contributions of great mathematicians go well beyond this field. Many of them were polymaths of one form or another, working in various disciplines, such as the sciences and philosophy.

Euclid is widely acclaimed as the "father of geometry," geometry being the basis of Greek mathematics.

A study of the history of mathematics is a study of the history of civilization. It can even be argued that the scientific revolution of the Renaissance came about because of the mathematical advances that allowed it. When Fibonacci (see pp. 80–81 and 85–87) introduced Hindu–Arabic numerals to Europe in the thirteenth century, they freed mathematics from the constraints of Roman numerals.

Mathematics did not advance at the same speed everywhere. Its progress has ebbed and flowed. Ideas have been discovered and lost, and then found again. The flow of knowledge hasn't been all in one direction, and modern mathematics takes ideas from many different places. However, we have much to be grateful to Arabic and Persian mathematicians for. They incorporated the best from the Greeks and Indians, before their knowledge was exported back to Europe and the cradle of the Renaissance.

In the modern world, mathematics has become ubiquitous. Of course, it has always been around us in many forms. But today it plays a far larger role in the everyday lives of normal people than ever. The degree of sophistication in a modern computer

| 1 | 2 | 3 | 4 | 5 | 6 | 7 | 8 | 9 | 0 |

Indian-Arabic numerals ca. 969 CE which developed into something approaching the numbers we use in everyday life.

means that there's far more invested in technological wizardry, and with that comes some remarkably sophisticated mathematics.

But you don't have to be a computer whiz or a math genius to appreciate the beauty of numbers. The more you become aware of mathematics, the more you see its influence in the world around you—you don't necessarily have to understand every last equation. Even the most sophisticated strands of mathematics, for example chaos theory, can be found in such everyday images as the wisps of smoke from a cigarette or the swirl of cream in your coffee.

The Nature of Mathematics

Mathematics is unique in that it can be both tangible, something you can touch, and yet also completely abstract. For example, addition can be demonstrated with something as tangible as a handful of pebbles, but at the same time the sum $2 + 2 = 4$ is a generalized statement that can be related to any object, or it can even be a completely abstract expression that has no physical embodiment.

Much of the history of mathematics has involved its development from the concrete to the abstract. For the ancient Greeks, mathematics was a very practical subject, with geometry (the study of shapes) as its basis. A variable (x or y, for example) was represented as a length, the square of that variable as an area, and its cube as a volume. However, such a pragmatic approach gave the Greeks issues when it came to dealing with ideas that fell outside this paradigm, such as negative numbers.

Over the course of the intervening millennia, mathematics has become more abstract in form, and therefore more flexible. But this doesn't mean that its applications are any less practical. Even when an idea is pursued on a purely theoretical basis, it can eventually find its way into everyday usage. A good example is Joseph Fourier, a French mathematician (1768–1830), who worked with infinite series of trigonometric functions. Yes, it's as complex as it sounds, and during his lifetime this subject was purely theoretical in nature, a mathematical puzzle to be solved, seemingly for its own sake. However, many years later the foundations he laid form the basis of analog–digital conversion, the technique that's used to turn analog sound waves into digital CDs.

The Language of Mathematics

$$\text{``}2x(3x - 4) = 6x^2 - 8x\text{''}$$

One of the most fascinating things about mathematics is that it's a universal language. Though there are many different tongues on the planet, there's one common form of mathematics. I have many exchange students in my classes, mostly from Europe and Asia. When they bring in their textbooks from their home country I can't understand a single word, but I do understand the mathematical symbols.

Amazingly, mathematics may be universal in the truest sense of the word. And it is for this reason that the Search for Extraterrestrial Intelligence opted for binary representations of π (see pp. 20–21) and prime numbers (see p. 16) to broadcast our presence to anyone who might be listening. The reason is that intelligent life on other planets would be unlikely to understand the word "hello" in any language. It's much more likely that they would have a concept of π, developed from working with circles; and, although

their main mathematical system may well be different from our base-ten system (like the decimal system we use, made up of tenths, tens, hundreds, and so on), they're likely to understand the concept of binary (on/off or day/night).

What Is Algebra?

The word "algebra" comes from a work by Al-Khwarizmi (see pp. 70–71) called *Hisab Al-Jabr w'Al-Muqabala*, where "Al-Jabr" was to become "algebra." And Al-Khwarizmi is considered by some to be the "father of algebra."

When we're talking about algebra in these pages, we're really talking about the elementary algebra taught in high schools around the world. Our focus is on the algebra that deals with arithmetic operations on numbers and variables; in other words, things that look like $3x + 5 = 9$. But, of course, algebra didn't spring into being fully formed: that sort of notation is relatively recent and stems from the seventeenth century and the work of René Descartes (see pp. 106–107).

"Mathematics knows no races or geographic boundaries; for mathematics, the cultural world is one country.**"**

—**David Hilbert**

The appearance of mathematical notation in its modern form coincided with the work of **René Descartes**.

The first stage of development was rhetorical algebra. This took the form of full sentences, and dominated until the third century. Today, this style of algebra remains the bane of most students because it's all about word problems. Rather than solve $3x + 5 = 9$, we would have to solve something as archaic as, "a quantity increased itself three times then increased five more equates to the value nine."

Then came syncopated algebra, where symbols and shorthand were introduced. The work of Diophantus (see pp. 54–55) is considered syncopated, as is the work of Brahmagupta (see pp. 64–67). This was an improvement, but still required a lot of extra work compared to what came next.

The final stage in algebra's development, at least as far as we're concerned, was symbolic algebra. This is the kind we know and love today. When we write $3x + 5 = 9$, x is our unknown and we can solve it by using the other information available. The question is truly theoretical, and doesn't need to be tied to a practical example.

ON NOTATION

The notation used throughout is standard; but it's worth noting that to avoid confusion the use of x has been reserved for variables, while the symbol • is used to denote multiplication.

For the sake of clarity, numerals and words are used where most appropriate. Likewise, units of measurement are usually metric without conversions, because it is the numbers that are important.

This notation came to its fruition with René Descartes, although it had been developed to a certain degree before him. Descartes' works are the earliest that a modern student could read without having real difficulty understanding the notation.

Alongside this, as we've already seen, algebra passed through several stages of increasing abstraction. During the Babylonian, Egyptian, and early Greek eras mathematics was geometric in nature, rendering the conception of zero and negative numbers absurd. Even when the algebra became syncopated, the dislike of negative numbers remained. Algebra then moved to a static equation-solving stage, and this is what you'll find within the rest of these pages.

1

The Basics of Algebra

To kick things off, we'll take a look at a few of the fundamentals. We'll start off by meeting some of the different types of numbers, including the perfect, the radical, the irrational, and everybody's favorite number: π. Then we'll move onto some of the ways in which you can solve the more basic algebraic equations, as well as unveiling some of the fascinating history behind it all.

What's in a Number? Part 1

A number is just a number, right? Well, not really. Numbers are much like people: they belong to different groups. Just like a high school will have the cool kids, the geeks, and so on, so do numbers. In fact, some numbers are squares, some are perfect, and one is even golden. Before we meet the golden and perfect numbers, let's look at the most basic of classifications for numbers.

Number Sets

So, you're a cave dweller, and you're entertaining yourself by counting rocks. This represents the most basic of number systems, the "natural" or "counting" numbers: 1, 2, 3, 4, 5, and so on. This number set worked well for many years and still does a great job for many things. To expand upon this number set, however, requires a great leap in thinking. We will add a single number to the natural numbers, zero (see box opposite), to create a new number set: the "whole" numbers. The whole number set is 0, 1, 2, 3, 4, 5, and so on.

The next number set, which includes the whole numbers, is the bane of many of our lives. This is the set of "negative" numbers. Where would commerce and banking be without negative numbers? Positive and negative whole numbers are called integers. The integer set is ..., –3, –2, –1, 0, 1, 2, 3, ... and can also be expressed as 0, ±1, ±2, ±3, ... and so on.

For the next set—the "rational" number set—you're out of your cave, agriculture has taken hold, and you're now raising chickens.

You wish to trade your chickens for a cow. So you meet with the cowherd and find out that the going rate for a cow is twenty chickens. You only have fifteen but you can use five of your brother Bob's to make up the difference. So, once you've butchered it, how much of the cow do you have to give Bob? Well, Bob gets $\frac{5}{20}$ of the cow, which simplifies to $\frac{1}{4}$, or a quarter.

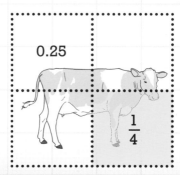

Rational (terminating or repeating decimals)

Integers
(0, ±1, ±2, ±3, ...)

Whole
(0, 1, 2, 3, ...)

Natural
(1, 2, 3, ...)

Irrationals (π, $\sqrt{2}$, ...)

While there is a pleasing relationship between integers and natural, whole, and rational numbers, the irrationals stand proudly apart.

These are fractions, or rational numbers. This set includes any number that can be expressed in the form $\frac{a}{b}$, where a and b are integers and b cannot equal zero—dividing by zero is a big no-no. Another way to express this is to say that rational numbers can be any "terminating" or "repeating" decimal. For example, $\frac{1}{4}$ is equal to 0.25, which is a "terminating" decimal. If, however, there was an exchange rate of nine chickens to a cow, and Bob had six chickens and you had three, Bob would get $\frac{6}{9}$ of a cow or $\frac{2}{3}$ or

0.66666..., which is a "repeating" decimal. Both of these types of decimal are rational.

So far the number sets we have met fit inside each other like Russian dolls, but the next stands apart from them. Numbers that cannot be expressed in the form of a fraction—in other words non-terminating and non-repeating decimals—are called "irrational" numbers. Two good examples are π (pi) and $\sqrt{2}$ (the square root of two); these are weird numbers because they continue forever, without repeating or ending.

ZERO THE HERO

Although we use zero on a daily basis, few of us ponder its significance. Zero is a vital part of our place-value system. Without it, 206 and 26 would look very similar indeed. Although this might seem obvious to us now, the theoretical leap required to develop a symbol that represents nothing is very impressive—and neither the ancient Greeks nor the Romans had a representation of zero.

The renowned Indian mathematician Brahmagupta (see pp. 64-67) was the author of the first text to treat zero as a number. It's sometimes said that you cannot ponder the infinite until you have pondered zero. In fact, this pondering of zero and the infinite is a big part of calculus—the nightmare of many university students. In essence, calculus is used in science, economics, and engineering to look at the infinitely large and infinitely small. So, not to put too fine a point on it, the appearance of zero was a huge moment in the history of mathematics.

What's in a Number? Part 2

Numbers have a varied social life, and they belong to different groups, like our chess clubs, gyms, or charities. On the previous pages we saw how numbers hang around in groups that fit within one another, like Russian dolls, and we also met the irrational numbers, which don't fit in; so here we'll look at a few other ways of grouping numbers.

Primes Versus Composites

Prime numbers are a subset, or smaller part, of the natural numbers. A prime is a natural number with exactly two distinct natural number divisors: one and itself. Or, to put it another way, a prime is a natural number that can be evenly divided only by one and itself. That is to say that if you divide a prime by any other natural number you will get a fraction or decimal. There are a couple of restrictions on primes: a negative number cannot be a prime; and one itself is not a prime.

Now, composite numbers represent the "opposite" of primes. A composite number is a natural number that has a positive divisor other than one and itself. This means that composite numbers represent all the natural numbers that are not prime numbers, except one. The number one is neither a prime number nor a composite number. There's always an odd one out.

Square Numbers

When we read $4^2 = 16$, we say "four squared equals sixteen." Ever wonder why we call it "squared"? Well, the Greeks were big on geometry and applied it to numbers as well. Sixteen is a square number because you can arrange sixteen dots to form a four-by-four square. In fact, sixteen is the fourth square number, or $n = 4$. Most of us are familiar with the square number set 1, 4, 9, 16, 25, and so on—it's the diagonal on a school multiplication grid—and the formula for finding a square number is n^2.

The Greeks applied geometry to numbers as well: Sixteen is a square number because you can arrange sixteen dots as a four-by-four square.

The first few square numbers (1, 4, 9, 16) arranged as dots.

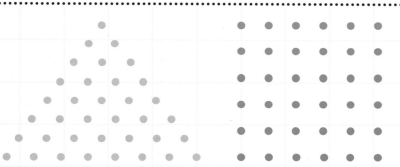

Thirty-six expressed as dots, showing that it is both a square number and a triangular number.

Triangular Numbers

A less well-known set is that of the triangular numbers: 1, 3, 6, 10, 15, 21, and so on. As with the square numbers, the triangular numbers get their name because they can be formed into triangles of dots.

It's interesting to note that there are some numbers that are both square and triangular. In fact, we've already met the first: it's the number one. After that we move onto thirty-six before we meet another number that can be drawn in dot form both as a triangle and a square, after that it's 1225, then 41616—the gaps become bigger as the numbers get higher.

The first nine square triangular numbers; by the ninth number we are already into the trillions.

1

3 6

1 2 2 5

4 1 6 1 6

1 4 1 3 7 2 1

4 8 0 2 4 9 0 0

1 6 3 1 4 3 2 8 8 1

5 5 4 2 0 6 9 3 0 5 6

1 8 8 2 6 7 2 1 3 1 0 2 5

Perfect Numbers

A "perfect" number is one for which the sum of all its proper divisors (whole number divisors) is equal to the number. This is best shown with an example: six is a perfect number because the proper divisors of six are one, two, and three. These numbers also total six. Perfect numbers are quite rare and really neat. The next perfect number is twenty-eight. The divisors of twenty-eight are one, two, four, seven, and fourteen. If we add these together then we get twenty-eight.

We don't see another perfect number until 496, and after that there isn't one until 8128.

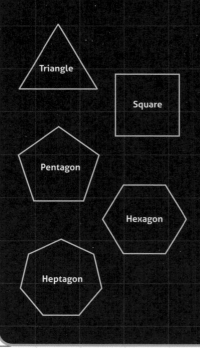

THE GEOMETRY OF NUMBERS

Like square and triangular numbers there are many geometric (or figurate) number sets that can be represented as geometric shapes, and the table below shows the first few along with the formulas that you can use to find them—just replace n with any number and you'll find the corresponding geometric number. Geometric numbers even exist in three dimensions; for example, there are "tetrahedral" numbers that are the sums of triangular numbers and form a pyramid with a triangular base.

6	**1 + 2 + 3 = 6**
28	**1 + 2 + 4 + 7 + 14 = 28**

Number Type	First Few Numbers	Formula
Triangular	1, 3, 6, 10, 15, …	$\frac{n(n+1)}{2}$
Square	1, 4, 9, 16, 25, …	n^2
Pentagonal	1, 5, 12, 22, 35, …	$\frac{(n)(3n-1)}{2}$
Hexagonal	1, 6, 15, 28, 45, …	$(n)(2n-1)$
Heptagonal	1, 7, 18, 34, 55, …	$\frac{(n)(5n-3)}{2}$

Backstage with Pi

Pi (π) is like a rock star. One Christmas my wife gave me a shirt with π on it. Whenever I wear it, strangers come up and say: "Cool shirt!" People like π. They like the idea of π; it connects them to mathematics beyond the mundane arithmetic of everyday life. For many, it's their first introduction to infinity. So here's a short history of π, its use, and significance.

What Is Pi?

Pi, or π, is defined as the ratio of the circumference of a circle to its diameter:

$$\pi = \frac{\text{circumference}}{\text{diameter}} = \frac{c}{d}$$

This often leads to confusion among people because they are also told that π is "irrational" which means that it cannot be expressed as a fraction. The thing we need to remember is that a fraction is $\frac{a}{b}$ where a and b are integers. But with π either the circumference or the diameter will be irrational. This is strange: it means, if you can write the value of the diameter, you cannot write the exact value of the circumference as a decimal, and vice versa.

The idea of π as a constant has been around for millennia. The Egyptians estimated it at $\frac{25}{8}$ (or 3.125) while the

Source	Year	Estimate
Rhind papyrus	ca. 1650 BCE	3.16045
Archimedes*	250 BCE	3.1418
Ptolemy	150 CE	3.14166
Brahmagupta	640 CE	3.1622 ($\sqrt{10}$)
Al-Khwarizmi	800 CE	3.1416
Fibonacci	1220 CE	3.141818

(*average of the bounds)

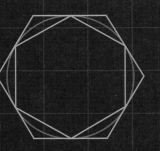

Archimedes' method for estimating π involved drawing regular ploygons inside and outside a circle, measuring them, and averaging their "bounds."

Mesopotamians gave it a value of $\sqrt{10}$ (or 3.162).

Archimedes was the first to study π in depth. By using polygons (many-sided shapes) both inside and outside the circle, and finding their perimeters, he was able to estimate for π between $\frac{223}{71}$ and $\frac{22}{7}$, which is where the common approximation of π as $\frac{22}{7}$ comes from. Since Archimedes' time the accuracy of π has been increasing, though some early estimates were better than those that followed (see table opposite). Now, thanks largely to the advent of computers, we know π to billions of digits.

Formulas for Pi
....

The symbol π (its modern mathematical meaning) was introduced by William Jones in 1706 in his book *Synopsis Palmariorum Mathesios*. However, π can also be represented as an infinite series of numbers. The fourteenth-century Indian mathematician and astronomer Madhava produced the series:

$$\frac{\pi}{4} = 1 - \frac{1}{3} + \frac{1}{5} - \frac{1}{7} + \frac{1}{9} \ldots$$

This can be used to estimate π, but it's slow. The eighteenth-century Swiss mathematician Leonhard Euler (see pp. 124–25) used the series:

$$\frac{\pi^2}{6} = 1 + \frac{1}{2^2} + \frac{1}{3^2} + \frac{1}{4^2} \ldots$$

While another interesting series was given by John Wallis (see box above), in 1656. It starts off with:

$$\frac{\pi}{2} = \frac{2}{1} \cdot \frac{2}{3} \cdot \frac{4}{3} \cdot \frac{4}{5} \cdot \frac{6}{5} \ldots$$

Without delving too deeply into the mathematics, these series show some of the many neat properties of π; and perhaps this is the reason for its enduring appeal. What's more, from the speedometer and odometer in your car to the calculation of the volume of every tin can, π's impact on everyday life can be felt everywhere.

Playing by the Rules

Imagine living in a place where people ignored the rules of the road—I could describe the nightmare of being stuck at a four-way stop in my hometown, but there aren't enough pages in this book. The point is, that without rules some people would drive on the right, some on the left, some would stop at red lights, some at green—it would be anarchy.

Law and Order

A similar structure and order is needed for mathematics. Without rules different people would get different answers to the same question. For example, let's look at how Sam and Mia solve the problem: 3 + 4 • 5.

Sam is a straight-ahead sort, so he adds three to four giving seven, then multiplies that by five to give thirty-five. Mia, on the other hand, does things differently: four times five is twenty, which added to three gives twenty-three. They get different answers and a fight ensues.

So who's right? Well, it's Mia. And here's why.

In mathematics there is a certain order in which mathematical operations must be done. There's a mnemonic to help: BEDMAS (sometimes BODMAS or PEDMAS). This translates as:

Brackets (or **P**arentheses)
Exponents (or **O**rders)
Division and **M**ultiplication
Addition and **S**ubtraction

You start with things inside brackets then move onto exponents (in 8^2, 2 is the exponent). Multiplication and division are all done at the same time, starting from the left and moving to the right. Then comes addition and subtraction—again, these are done at the same time, starting from the left and moving to the right. Higher functions, such as logarithms and trigonometric functions, which we'll look at later in the book, happen at the exponent level.

The levels make sense if you think about the operations that are performed. Addition is the the first operation we learn. Multiplication is really just successive additions; that is to say, "two times five" is really two added to itself five times:

$2 \cdot 5 = 2 + 2 + 2 + 2 + 2$

Next, an exponent represents successive multiplications, for example:

$$2^5 = 2 \cdot 2 \cdot 2 \cdot 2 \cdot 2$$

So, each operation builds on the previous one as we move through the order of operations.

Nested Brackets

Nested brackets (a set of brackets inside another set) sometimes cause problems when trying to follow BEDMAS. So let's look at an example:

$9 + 3(8 - 2(6 - 5))$

To simplify this we evaluate from the inside out. Therefore, we first subtract $(6 - 5)$ to get 1. Now the expression is:

$9 + 3(8 - 2(1))$

And $2(1)$ is really just 2, which gives:

$9 + 3(8 - 2)$

Next we evaluate our final bracket, to get 6 and the expression is:

$9 + 3(6)$

This becomes $9 + 18$ which equals 27.

Grouping

Another problem people confront is grouping, which is often implied and can cause confusion. As an example $x \cdot x - 3$ is not the same as $x \cdot (x - 3)$. The former is really $x^2 - 3$ while the latter is $x^2 - 3x$.

So when someone writes $\frac{1}{2}x$ do they mean "one half times x" or do they mean "one divided by two xs"? If x had a value of 10 the first result would be 5 while the second would be 0.05—a big difference. This is where grouping helps; if you want to say "half times x" you could write $\frac{1}{2}x$ to avoid confusion. (We've used a vinculum, a fancy term for the horizontal dividing line, throughout, in order to avoid confusion.)

Lastly, with the division line there is one assumed rule—everything above or below a horizontal division line is inside assumed brackets. So the expression $\frac{x+1}{x-3}$ could be written out explicitly as $\frac{(x+1)}{(x-3)}$.

This is by no means a definitive list of the order of operations. In computer science there is a much larger list of operations and the corresponding order in which they are carried out. In mathematics we also have many more operations that can be performed, such as factorials (see pp. 112–13).

Expressions, Equalities, and Inequalities

Before we look too closely at equalities and inequalities, we need to understand the terminology. An expression is a collection of numbers and variables that can sometimes be simplified—it has neither an equals nor an inequality (ie, "doesn't equal") sign in it. An equation always includes an equals sign, while an inequation has an inequality sign where the equals sign would normally be.

One example of an expression is:

$$\frac{(3x - 4) + 5}{5x}$$

While an example of an equation is:

$$3x - 5 = 13$$

And an example of an inequation is:

$$3(x + 2) \le 2x + 5$$

When it comes to inequalities: > means "greater than"; ≥ means "greater than or equal to"; < means "less than"; and ≤ means "less than or equal to."

Inequalities might seem odd, but we meet them all the time in everyday life; for example, when we're thinking about maximums and minimums.

Inequalities on a Number Line

Because inequalities have an infinite number of solutions, we often represent them on a number line. If we put $x \le 3$ on a number line there would be a dot at three and the number line to the left would be shaded (see below). This represents all the values that make the inequality true. If we put $x > -2$ on a number line it would be a hollow dot at minus two, and the number line to the right would be shaded. A solid dot is used when the number forms part of the solution, (in other words, when we use ≤ or ≥); while a hollow dot is used when the number is not part of the solution, (in other words, when we use < or >).

For example, $3x - 5$ is an expression; and it's one that we can't really do much with. Now, inserting an equals

-9 -8 -7 -6 -5 -4 -3 -2 -1 0 1 2 3 4 5 6 7 8 9 sign with something on the other side makes it an equation. To extend the example above, $3x - 5 = 13$ is an equation, so we can set about finding its solution, which is $x = 6$. Meanwhile, $3x - 5 > 13$ is an inequation which can also be solved. The solution is $x > 6$.

So, the equality generates a discrete (or definitive) solution: x is six and nothing else. On the other hand, the inequality generates a range of solutions: x could be seven, eight, or 6.000001; in fact, it is anything greater than six. For an inequality there are an infinite number of solutions, and this is often expressed on a number line by shading to the right of the six (see above). Notice the six has a hollow circle above it, because a hollow dot signifies that it's not included in the solution.

Changing Direction
....

There's a common stumbling block with inequations, and the next example demonstrates this. Suppose that Sam and Dave made a deal with the Devil: in exchange for fame, one of them will have to give up his soul. The Devil gives them a task and whoever gets it right saves his soul. The task is to solve $-3x > 15$.

Sam and Dave both solve it, but in different ways. Sam divides both sides by -3 and gets $x > -5$. Dave, however, moves the $-3x$ to the right and the 15 over to the left to get $-15 > 3x$. Then Dave divides by 3 to give $-5 > x$. So Sam has numbers greater than -5 while Dave has numbers less than -5.

If Sam is right then -4 should work; but it doesn't, and we get $12 > 15$. For Dave -6 should work. When we plug it in we get $18 > 15$, which is true. This demonstrates an important and often forgotten rule relating to inequalities: when multiplying or dividing by a negative number the direction of the inequality must switch.

GIVE US A SIGN

Although many of our mathematical symbols and notations are standardized today, many are not. In the past even fewer were standardized. The = sign was introduced in 1557 by Robert Recorde, a Welsh doctor and mathematician. The symbols > and < were first introduced in a book by Thomas Harriot, an English mathematician, in 1631, some ten years after his death; although his editor is credited for the notation. Then, over a hundred years later, in 1734, the symbols ≥ and ≤ were introduced by the French mathematician Pierre Bouguer.

The Power of Polynomials

Polynomials are important because they are used to model real-world problems. For example, degree-one polynomials (lines) are used in business for optimization problems (finding the best solutions), while degree-two polynomials (quadratics) are used to model problems including those involving gravity. Higher-degree polynomials are often used to model complex systems such as the economy.

Introducing the Terminology

- A **polynomial** is a collection of terms.

- A **term** is a collection of variables raised to exponents and multiplied by a coefficient.

- In the example $3x^2$, 3 is the **coefficient**, x is the **variable**, and ^{2} is the **exponent**.

Another example of a term would be $5xy^3$; where 5 is the coefficient, x and y are the variables, and 1 and 3 are the exponents. Note that although there is no exponent on the x it's implied that there is a 1 there.

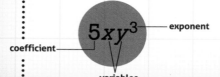

coefficient —
exponent
variables

There are a few ground rules for polynomials: the exponents must be whole numbers, not fractions, nor negative or irrational numbers. So terms like $4x^{\frac{1}{2}}$, $2\sqrt{x}$, and $\frac{5}{x^2}$ cannot be in a polynomial because the exponents on x are not whole numbers.

Naming Polynomials

Polynomials can be defined by the number of terms in them: a polynomial of one term is called a monomial, whereas a binomial has two terms; a trinomial has three; and a polynomial has more than three ("poly" meaning "many"). Meanwhile, the degree of a polynomial is based on the term with the highest sum of exponents. For example, the expression $3x^2 - 4x + 5$ is a trinomial of degree two, because the highest exponent is a ^{2}.

Polynomials are very important in algebra. Polynomials of degree one (or linear functions) have been used to solve problems going back centuries. Degree-two polynomials (or quadratic functions) were studied by ancient Babylonian, Greek, Indian, and Arab mathematicians and are used in most branches of science, engineering, mathematics, and economics. The Babylonians used tables of squares (quadratics) to solve multiplication problems. They used the formula:

$$ab = \frac{(a + b)^2 - (a - b)^2}{4}$$

This allowed them to look up sums and differences in the table and divide by four. As an example $12 \cdot 8$ would be:

$$\frac{20^2 - 4^2}{4}$$

In this case 20^2 and 4^2 would be found in a table, and plugged in as:

$$\frac{400 - 16}{4}$$

This can be simplified to:

$$\frac{384}{4}$$

Which reduces to 96.

A Brief History of Polynomials

. . . .

Polynomials have been studied for many years; as we've mentioned earlier, solutions to quadratics (degree-two polynomials) stretch as far back as the ancient Babylonians.

NIELS ABEL

Born in Norway in 1802, Niels Henrik Abel spent a good portion of his life in poverty; but luckily for him his mathematics teacher recognized his talent and helped support him through his higher education. He graduated from university in 1822, and just two years later he published his most notable work, which proved that there is no general solution to a polynomial equation of degree five.

In recognition of his work, the Norwegian government sponsors the Abel Prize, which some call the "Nobel Prize for mathematics," as there is no Nobel Prize in the field.

The ancient Greek mathematician Euclid (see pp. 44–45) solved quadratics with a purely geometrical approach around 300 BCE; but it was around a thousand years later when the Indian mathematician Brahmagupta (see pp. 64–67) gave an almost modern approach to solving quadratics.

Later, in sixteenth-century Italy, the cubic and quartic equations (polynomials of degrees three and four) were the subject of some serious mathematical one-upmanship. Then, in 1824, Niels Abel (see above) proved that there is no general solution to polynomial equations of degree five.

Talking Trig

The meaning of trigonometry comes from two Greek words, *trigonon* (triangle) and *metron* (measure)—to measure the triangle. Its development spans all cultures, mostly due to the connection between trigonometry, astronomy, and navigation, and it still has many uses today—for example, in surveying and mapping.

The sextant, invented in the early eighteenth century, is used in conjuction with the principles of trigonometry to aid nautical navigation.

The Babylonians

Around three millennia ago, the ancient Babylonians had a form of trigonometry, and it's from them that we take the idea of 360° in a circle. They also gave us sixty minutes in a degree and sixty seconds in a minute. That is to say, when you have 20.5° it can be written 20° 30', which reads "twenty degrees thirty minutes." It's also the route of having sixty minutes in an hour and sixty seconds in a minute. All because the Babylonian number system was base-sixty (sexagesimal) and six sixties were a full circle.

20° and 30 minutes (20.5°)

The Greeks

The Greeks also worked with advanced trigonometry. Euclid (see pp. 44–45) and Archimedes (see pp. 46–47) developed theorems, albeit via geometry, which have trigonometric equivalents.

It should be noted that the trigonometry of the ancient Greek world looked different from the trigonometry of today, as theirs was based on "chords" of circles—lines from edge to edge.

The first trigonometric table is thought to have been compiled by the second-century BCE mathematician and astronomer, Hipparchus of Nicaea, and some call him the "father of trigonometry." The table was developed to aid in solving triangles, and Hipparchus is also credited with introducing the Greeks to the idea of 360° in a circle.

Somewhat later, Menelaus of Alexandria (ca. 70–130 CE) wrote on spherical trigonometry, and the astronomer and geographer Ptolemy (ca. 85–165 CE) furthered Hipparchus's work in his thirteen-volume *Almagest*.

This title page (right) from a 1611 astronomy publication features a number of famous figures, including Hipparchus (top).

SIMILAR TRIANGLES

"Similar" triangles are those that have the same three angles in common. Though the triangles can be of varying sizes, the ratio of their corresponding sides will remain constant. Therefore, if we know that A is twice as long as a, then the value of B will be twice b, and C will be twice c.

$C = 12$

$A = 8$

o

$c = 6$

$a = 4$

$B = 6$

$b = 3$

Indians and Persians

The Indian mathematician Aryabhata (476–550 CE) developed the ratios for sine and cosine (see box) that most closely resemble our modern forms. His work also contains the earliest surviving sine tables. In the seventh century the Indian mathematician Bhaskara produced a fairly accurate formula (for use with radians, a unit of measurement for angles, not degrees) to calculate the sine of x without a table:

$$\sin x \approx \frac{16x\,(\pi - x)}{5\pi^2 - 4x(\pi - x)}, \; (0 \leq x \leq \tfrac{\pi}{2})$$

These ideas migrated west through Persia. Al-Khwarizmi (see pp. 70–71) produced trigonometric tables for sine, cosine, and tangent in the ninth century. A century later, Islamic mathematicians were using all six ratios and had tables for quarter-degree increments accurate to eight decimals. In the eleventh century Al-Jayyani, who was born in Córdoba, Spain, produced work containing formulas for right-angled triangles which likely influenced European mathematics.

Trigonometry Today

Today there are a huge number of practical applications for trigonometry. As well as surveying and mapping, which we've already mentioned, it's put to use in navigation. For example, both the traditional sextant that sailors used to gauge their position on the world's oceans and the more modern invention of satellite navigation systems use trigonometry. It's also used on a more theoretical basis to model financial markets among many other areas.

This page from an early seventeenth-century book show the values of trigonomic functions.

SIX OF A KIND

You may already be familiar with the six ratios used in trigonometry that link the angles and the length of the sides of a right-angled triangle, if only because you have seen them on a scientific calculator. Put simply, if you know certain angles and side lengths, you can use these ratios to determine the values of those you don't. All you need to know is which ratio to use:

sine θ = opposite ÷ hypotenuse

cosine θ = adjacent ÷ hypotenuse

tangent θ = opposite ÷ adjacent

cosecant θ = hypotenuse ÷ opposite (note this is the reciprocal of sine)

secant θ = hypotenuse ÷ adjacent (note this is the reciprocal of cosine)

cotangent θ = adjacent ÷ opposite (note this is the reciprocal of tangent)

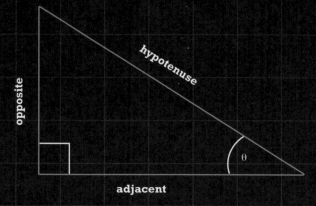

Classical Math

When we think about the history of mathematics it's normal to think about ancient Greece. After all, we take many of our mathematical symbols from the Greek alphabet: we've already met π and φ (the golden ratio). And if you ask anyone to name a famous mathematician, the chances are they'll choose Pythagoras, Archimedes, or some other long-dead Greek. To a certain extent this familiarity exposes a Western bias in our thinking, but it does make ancient Greece a good place to start our journey into algebra.

Pythagoras

Pythagoras is often the first historical mathematician people are introduced to. Up until Pythagoras, mathematics is really quite nameless. Throughout most of our early years we learn about arithmetic, and it's not until we reach the Pythagorean theorem that we get a chance to put a face to the numbers. As it turns out Pythagoras is quite a good person to start with as he had a profound influence on mathematics.

Pure Ability

Pythagoras of Samos was born about 570 BCE on the Greek island of that name, in the Aegean near Turkey. He's sometimes described as the first pure mathematician, studying mathematics as a theoretical pursuit, not a practically applied discipline. This is important because making the intellectual leap from five apples, five people, and so on, to the abstract number five was a significant event.

Unlike many other historical figures, there are no primary sources for Pythagoras. Either his writings have been destroyed or Pythagoras did not write but rather had students record his thoughts. Therefore all of our accounts of Pythagoras come from other sources, some that seem reasonable and others fantastic.

The Life of Pythagoras

What we know of the life of Pythagoras is quite detailed considering that we have to go back two and a half millennia. He spent his youth in Samos but traveled extensively with his father, Mnesarchus, a merchant from Tyre. Pythagoras visited Thales, a Greek philosopher, scientist, mathematician, and engineer by trade, in Miletus and attended lectures given by Anaximander, a student of Thales.

Pythagoras also traveled to Egypt, and during a war between Egypt and Persia, he was captured and taken to Babylon. Around 520 BCE, Pythagoras returned to Samos. Shortly thereafter he went to southern Italy and founded his Pythagorean school at Croton.

Pythagoras's group, the Pythagoreans, was a bit of a cult, mixing religion and mathematics.

Pythagoras's group, the Pythagoreans, was a bit of a cult, mixing religion and mathematics. The group at Croton was in some ways a school, a monastery, and a commune. The group consisted of two parts. The first, called the mathematikoi, lived with and were taught by Pythagoras. These members were required to live an ethical life, practice pacifism and study the "true nature of reality"—mathematics or numbers. The second part was the akousmatikoi, who lived in their own house and attended the school during the day.

However, it wasn't all mathematics; the Pythagoreans also believed in the transmigration of souls and reincarnation, while one of the stranger rules the Pythagoreans followed was the abstinence of eating beans.

Pythagoras and Mathematics
· · · · · · · · · · · · · · · · · · · ·

Mathematically, the Pythagoreans are known for many things: the Pythagorean theorem, the mathematics of music, and the discovery of the square root of two. Pythagoras himself might not have been responsible for all these ideas; the members of the group no doubt contributed significantly. Given the secrecy surrounding the Pythagoreans, it's hard to determine what work was Pythagoras's and what wasn't. (The Pythagorean theorem is discussed in detail on the following pages and the square root of two is mentioned in the section on irrational numbers.)

In 508 BCE the Pythagorean society was attacked by Cylon, a noble from Croton. Pythagoras escaped to Metapontium and died around eight years later.

PYTHAGORAS AND MUSIC

Pythagoras and the Pythagoreans were very interested in music, being musicians as well as mathematicians. One story goes that Pythagoras was passing a blacksmith when he heard harmonious notes emanating from the shop. Upon inspection Pythagoras realized the notes were related to the size of the tool. Using simple fractions Pythagoras produced the notes that we know today. If you have a string and pluck it to produce a "C" then shorten the string to half its length and pluck again you will hear a "C," one octave up. A string shortened to half its length produces a frequency that is twice as big, in other words an octave higher.

Note	String length
C	1
D	$\frac{8}{9}$
E	$\frac{4}{5}$
F	$\frac{3}{4}$
G	$\frac{2}{3}$
A	$\frac{3}{5}$
B	$\frac{8}{15}$
C	$\frac{1}{2}$

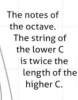

The notes of the octave. The string of the lower C is twice the length of the higher C.

Pythagorean Theory

The Pythagorean theorem is one of the best-known pieces of mathematics—it's one of the those indispensable bits of math that get drilled into kids in every school. The theorem also has many everyday applications.

A Simple Proof

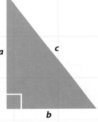

The Pythagorean theorem states that given a right-angled triangle (below) the sum of squares of the two shorter sides equals the square of the longer side, or as the formula: $a^2 + b^2 = c^2$.

Though the formula is named after Pythagoras, the theorem was actually known to the Babylonians and Indians long before his time. However, it is thought that Pythagoras, or perhaps one of the Pythagoreans, was the first to construct a proof of the theorem.

So here is one of the many ways that it can be proved. The area of the large square is:

$(a + b)^2$
or
$(a + b)(a + b)$

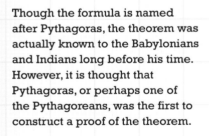

After we "foil" this out (whenever we multiply two binomials we foil) and collect the like terms we have:

$a^2 + 2ab + b^2$

The area of the large square can also be determined by adding the areas of the four triangles and the smaller square of side length c. The area of the slanted square is c^2, and the area of one of the triangles is $\frac{1}{2}ab$. So the sum of the four triangles and the slanted square is:

$4 \cdot \frac{1}{2}ab + c^2$

This simplifies to $2ab + c^2$.

Now, since we are talking about the same square, the areas must be equal, therefore:

$a^2 + 2ab + b^2 = 2ab + c^2$

The $2ab$ that appears on both sides of the equation cancels itself out leaving:

$a^2 + b^2 = c^2$

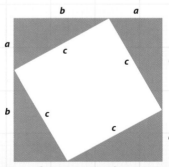

A Common Use of the Theorem

....

In construction, the Pythagorean theorem is used to check how square a corner is. A quick way to see if a corner is square (90°) is to measure down one wall three meters, down the other wall four meters, then measure the diagonal between these two points. If the diagonal is not five meters, then the walls are not square. You can extend down both walls by multiples of three and four to gain more accuracy.

Pythagorean Triples

....

The following example, using three, four, and five for the sides of a right-angled triangle, is called a "Pythagorean triple." There are many Pythagorean triples, including any multiple of three, four, and five; in fact, the aspect ratios (the ratio of length to height) of both regular (4:3 ratio) and widescreen (16:9 ratio) televisions are both part of Pythagorean triples. Here's the formula for determining them.

Given two whole numbers, n and m, such that n is greater than m:

$a = n^2 - m^2$
$b = 2nm$
$c = n^2 + m^2$

So if $n = 2$ and $m = 1$ then:

$a = 2^2 - 1^2 = 4 - 1 = 3$
$b = 2 \cdot 2 \cdot 1 = 4$
$c = 2^2 + 1^2 = 4 + 1 = 5$

Using different values of n and m you can create as many Pythagorean triples as you like.

The 3D Experience

....

One more interesting thing about the Pythagorean theorem is that it can be generalized to other dimensions. For those of us in the real world that means three dimensions.

As an example, say that Suzie is collecting little nick-nacks and plans to use a shoe box to store them. She sees a novelty pencil in a store, but needs to know if it will fit into the box before she buys it. Okay, so a normal person would just get on and buy it, but hey, don't knock Suzie—she's sensitive.

Luckily for us, Suzie knows the dimensions of the box: it's 18 cm wide, 28 cm long, and 11 cm high. She then applies the theorem to three dimensions or $a^2 + b^2 + c^2 = d^2$, where a, b, and c are the width, length, and height and d is the diagonal. So Suzie has:

$18^2 + 28^2 + 11^2 = d^2$

This becomes:

$324 + 784 + 121 = d^2$ or $1229 = d^2$

Taking the square root of both sides we find that Suzie can get a 35cm pencil.

Radical Mathematics

"Radicals," or more specifically square roots, have been known for centuries. The Rhind Papyrus (see p. 59) makes reference to square roots as long ago as ca. 1650 BCE, but this is not surprising given that square roots are tied to the areas and diagonals of squares and rectangles; so temple construction would have required an understanding of them. In modern times they've all sorts of uses, such as allowing electrical engineers to calculate the average power dissipation in a circuit.

Getting to the Root of Two

The square root of two ($\sqrt{2}$) was a big deal to the Pythagoreans (see pp. 36–37). The discovery that the square root of two was irrational really bothered them. To the Pythagoreans the world was all about numbers, and that meant rational numbers. The idea that a number could not be expressed as a fraction was inconceivable.

Legend has it that Hippasus of Metapontum, a disciple of Pythagoras, produced a proof of the irrationality of the square root of two. Because the Pythagoreans could not accept this, Hippasus was sentenced to death by drowning. An alternative story suggests that his discovery was made at sea, so he was simply thrown overboard. Who knows for sure? Maybe they just

expelled him from the group, but it makes for a good story, and illustrates just how irrational people can get about numbers.

Another Famous Greek

Archimedes (see pp. 46–47) had a very accurate estimate of the square root of three ($\sqrt{3}$), which he used in *Measurement of a Circle*, the same text in which he outlined his estimate for π.

Archimedes' estimate for $\sqrt{3}$ was $\frac{265}{153} < \sqrt{3} < \frac{135}{780}$, or expressed in decimals $1.7320261 < \sqrt{3} < 1.7320512$. It's worth noting that the second figure is out by only 0.0000004, which is very close considering that Archimedes didn't have a calculator or a base-ten system to work with— multiplication and division using Greek numerals was very difficult. Some mathematical historians suggest that Archimedes used the Babylonian method.

The Babylonian method is an elegant iterative (repetitive) formula. Given $x_0 \approx \sqrt{S}$ the estimate root is found using:

$$x_{n+1} = \frac{1}{2}\left(x_n + \frac{S}{x_n}\right)$$

As an example we'll estimate the square root of three ($\sqrt{3}$). Note that $\sqrt{3}$ on a calculator is 1.732050808. First we need a starting estimate: x_0. As we know the square root of four is two, we'll start there. It's too large, but the formula will help us work our way down to the right value. Substituting x_0 for x_n makes x_{n+1} equal to x_1.

$$x_1 = \frac{1}{2}\left(x_0 + \frac{S}{x_0}\right)$$

where $x_0 = 2$ and $S = 3$ (for $\sqrt{3}$)

$$x_1 = \frac{1}{2}\left(2 + \frac{3}{2}\right) = 1.75$$

Note that we already have the first two digits correct and we have a value for x_1, a better estimate for $\sqrt{3}$. Using $x_1 = 1.75$ we can now apply the formula again and get an even better estimate:

$$x_2 = \frac{1}{2}\left(x_1 + \frac{S}{x_1}\right)$$

Where $x_1 = 1.75$ and $S = 3$ (for $\sqrt{3}$)

$$x_2 = \frac{1}{2}\left(1.75 + \frac{3}{1.75}\right) = 1.7321$$

We've now found the first four digits of $\sqrt{3}$, and if we wanted we could continue this process to arrive at better and better estimates for the square root of three.

HERON OF ALEXANDRIA

Along with his method of determining square roots, Heron of Alexandria (ca. 10-70 CE) provided a neat way of finding the area of a non-right-angled triangle, using the formula:

$$\text{area} = \sqrt{s(s - a)(s - b)(s - c)}$$

where: $s = \dfrac{a + b + c}{2}$

or in another form:

$$\sqrt{(a + b + c)(a + b - c)(b + c - a)(c + a - b)}$$

Here a, b, and c are the sides of the triangle, and s is its semi perimeter (half its perimeter). Though this formula looks ugly it's really quite nice. Moreover, because of the difficulties in doing calculations in ancient Greece, Heron's formula offered a much easier solution.

Plato

Even if many people don't know exactly what he did, the name of Plato is familiar to most people in the Western world. He is one of the big guns of philosophy, but is also renowned in mathematical circles, not so much for any advances, but more for his attitude toward the subject. Plato was a torch bearer of mathematics; and through him the mathematics of Pythagoras and his followers was passed on to Euclid (see pp. 44–45) and Archimedes (see pp. 46–47).

War Child

Plato was born in Athens ca. 427 BCE. The son of wealthy parents, he was afforded a solid education. During his youth it's almost certain that Plato was a follower of Socrates, as Plato makes mention of him in his *Dialogues*. In fact, the arrest, trial, and execution of Socrates had a significant impact on Plato, and after the execution in 399 BCE he left Greece for Egypt, Sicily, and Italy.

When in Italy, Plato learned of the works of the Pythagoreans, and from these developed his ideas on reality.

The Pythagoreans are considered the first group to study mathematics as an intellectual pursuit, helping to separate the world of mathematics from the "real world," and this affected Plato greatly.

Plato considered mathematical objects to be perfect forms, which cannot be created in the real world. In *Phaedo*, Plato talks of objects in the real world trying to be like their perfect forms. As an example, a line in mathematics has length but no width, therefore it's impossible to draw a true line in reality, because a line on a page requires some width in order to be seen. A perfect line also continues forever, which is just

MAJOR WORKS

THE REPUBLIC: Plato's best-known work deals with the ideal society, ideal rulers, and forms of government. Plato also talks of imperfect representations of perfect mathematical objects.

PHAEDO: Describes Socrates' death and discusses the afterlife, giving four arguments for the soul's immortality. It also talks about perfect forms and imperfect representations of them.

TIMAEUS: Contains a construction that gives the Platonic solids the characteristics of the elements earth, fire, air, and water as well as the universe (see pp. 42-43).

> "Those who have a natural talent for calculation are generally quick-witted at every other kind of knowledge; and even the dull, if they have had an arithmetical training, always become much quicker than they would have been." —*The Republic*

as impossible to draw. Of course, we can draw lines with arrows on them to represent the infinite direction, but this is just a crude representation.

Plato and the Academy

Plato returned to Athens in 387 BCE and founded the academy where he worked until his death in 347 BCE. The academy was devoted to research into philosophy, science, and mathematics.

Though very fond of mathematics, Plato did not advance mathematical thought, though the ideas of Pythagoras continued through him, and his reverence of mathematics extended to his students. This makes Plato a very important link in the chain of Greek mathematics. In fact, in *The Republic* Plato states that one must study the five mathematical disciplines: arithmetic, plane geometry, solid geometry, astronomy, and music, before moving on to study philosophy.

Actually, there is some mathematics attached to Plato: the Platonic solids. Though they are named after him these five shapes with neat properties—the tetrahedron, cube, octahedron, dodecahedron, and icosahedron— were known to many before Plato's name was attached to them.

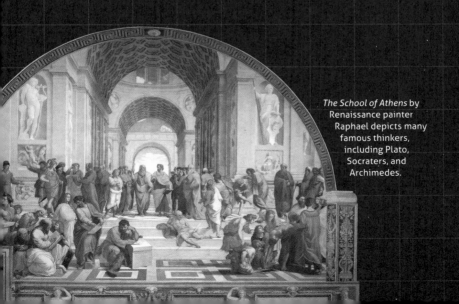

The School of Athens by Renaissance painter Raphael depicts many famous thinkers, including Plato, Socraters, and Archimedes.

Platonic Solids

The Platonic solids, as well as the Archimedean solids shown on pages 48–9, are nice curiosities of three-dimensional geometry. They can be seen in all sorts of interesting places, from such everyday objects as dice, to the shapes of molecules (methane is a tetrahedron) and even viruses (the herpes virus is an icosahedron).

The five Platonic solids, the tetrahedron, hexahedron (cube), octahedron, dodecahedron, and icosahedron, are all regular. This familiar-looking set of shapes is named after Plato (see pp. 40–41), though it's very unlikely that he discovered them.

Sources suggest that the Pythagoreans (see pp. 36–37) were aware of the regular tetrahedron, hexahedron, and dodecahedron. However, it's likely that credit for the discovery of the regular octahedron and the regular icosahedron belongs to Theaetetus (417–369 BCE), a Greek mathematician who studied under Plato and is the central character in two of

Plato's Dialogues. This idea is also supported in Book XIII of Euclid's *Elements*.

It's also believed that Theaetetus gave the first proof that there are only five such shapes, at least in the case of three-dimensional objects.

Properties of the Platonic Solids

First it's necessary to define what makes a Platonic solid. To be a Platonic solid an object must have congruent (equal) faces; these faces must only intersect at the edges; and the same number of faces must meet at each vertex (or point). What this really means is that no matter which

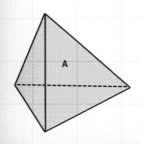

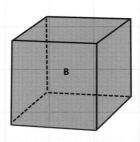

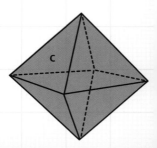

With the Platonic solids there is also a relationship between the faces, edges, and vertices. Looking at the tetrahedron, it has four faces, four vertices (where the sides come together at a point), and six edges. The cube has six faces, eight vertices, and twelve edges. Look at the final column and you'll see a neat symmetry to the relationships between the faces, vertices, and edges of all five Platonic solids.

Polyhedra	Faces (F)	Vertices (V)	Edges (E)	F + V - E
Tetrahedron	4	4	6	4 + 4 - 6 = 2
Hexahedron	6	8	12	6 + 8 - 12 = 2
Octahedron	8	6	12	8 + 6 - 12 = 2
Dodecahedron	12	20	30	12 + 20 - 30 = 2
Icosahedron	20	12	30	20 + 12 - 30 = 2

face is down (or at the bottom) the shape will look the same. This makes the Platonic solids fair, and that's why they're used as dice. The regular hexahedron (cube) is the common six-sided dice, while the others are used in role-playing games and wargames. The interesting relationship between the faces, vertices, and edges is shown in the table above.

Meanwhile, Plato also attributed some less mathematically sound, but still fascinating, attributes to the Platonic solids. He married them to the classical elements of the ancient world, making the tetrahedron fire, the hexahedron earth, the octahedron air, and the icosahedron water. The fifth Platonic solid, the dodecahedron was given the role of arranging the constellations.

These Platonic shapes were peculiar among geometrical figures in that they were the only known shapes that could be constructed whose faces, sides, and angles were all identical to one another.

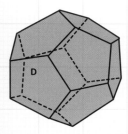

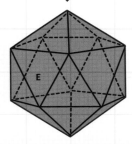

A= the tetrahedron (pyramid with a triangular base)

B= hexahedron (cube)

C= octahedron (8-sided regular solid)

D= dodecahedron (12-sides)

E= icosahedron (20-sides)

Euclid

Euclid of Alexandria is widely acclaimed as the father of geometry. His book, or collection of books, the *Elements* was the authoritative text on geometry for over two thousand years and has been called the most successful textbook in the history of mathematics. The geometry we learn in school is Euclidian geometry. In fact, until the early nineteenth century, it was the only geometry. Euclidian geometry is one of the most enjoyable branches of mathematics to learn and teach.

Euclid's Life

For someone who wrote one of the most influential books in mathematics, surprisingly little is known about Euclid. He was born around 325 BCE, though we don't know where. There is some suggestion that Euclid attended Plato's academy (see pp. 40–41), though most likely after the death of Plato. And he is said to have been active and teaching in Alexandria during the reign of Ptolemy I, one of Alexander the Great's generals, who ruled Egypt 323–283 BCE.

So little is known about Euclid that some speculate about whether he really existed or not. Three options have been presented: that Euclid did exist and wrote the books himself; that Euclid was a leader of a group of mathematicians who collectively wrote as Euclid (much like Pythagoras and the Pythagoreans); or that Euclid didn't exist but was a group of mathematicians writing as Euclid. Assuming that Euclid was alive in the first place—and I do—he is thought to have died around 265 BCE.

Elements of Geometry

Elements is Euclid's big work; a collection of thirteen books dealing with geometry and number theory. A common misconception is that *Elements* is a book of Euclid's own discoveries; but this isn't true, as most of the mathematics in it were known before Euclid's time. Euclid's real accomplishment was to collect and organize the information,

> "There is no royal road to geometry."
>
> —Euclid

and provide proofs for many of the ideas—thereby setting mathematics on a more rigorous path than before.

Books I to VI of *Elements* deal with plane geometry—the geometry we learn in school. Books I and II deal with triangles, squares, rectangles, parallelograms, and parallels. Book I also includes the Pythagorean theorem (see pp. 36–37). Book III describes the properties of circles, while Book IV deals with problems involving circles. Book V looks at commensurable and incommensurable magnitudes—mostly lines. (A commensurable magnitude being where two lengths form a ratio that is rational; while incommensurable means that the ratio is irrational.) Applications of the results from Book V are dealt with in Book VI.

Books VII to IX deal with number theory. Book VII includes the Euclidian algorithm for finding the greatest common factor of a pair of numbers. It also discusses primes and divisibility. Book VIII looks at the geometric progression of numbers.

These pages appeared in the first ever printed edition of Euclid's *Elements,* produced by the German Erhard Ratholt in 1482.

Among other things, Book IX looks at sums of geometric series and perfect numbers (see p. 18). While Book X again looks at irrational numbers—something that gave Pythagoras a real headache.

Books XI through to XIII deal with three-dimensional geometry, while book XII looks at the areas and volumes of spheres, cones, cylinders and pyramids. While Book XIII, the last book, deals with the properties of the five regular polyhedra (the Platonic solids).

OTHER WORKS

DATA: Deals with the properties of figures which can be deduced when other properties are given.

ON DIVISION OF FIGURES: Looks at dividing figures into two or more equal parts.

CATOPTRICS: A work on the mathematical theory of mirrors.

PHAENOMENA: Deals with spherical astronomy.

OPTICS: Covers the mathematics of perspective.

CONICS (LOST): A work on the conic sections (see pp. 74-75).

PSEUDARIA OR BOOK OF FALLACIES (LOST): A text on errors in logic.

Archimedes

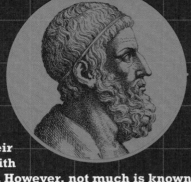

If Euclid is the geometry guy then Archimedes is the all-round guy. In fact, when mathematicians rate their greatest, Archimedes is up there with Newton and Gauss (see pp. 128–29). However, not much is known about the personal life of Archimedes. He was born in Syracuse, on Sicily in 287 BCE, and he spent most of his life there—although there is a belief that he traveled to Alexandria and spent some time at its famous library. When in Egypt it's likely that Archimedes studied with the successors of Euclid, and in the preface of *On Spirals*, he makes reference to his friends in Alexandria, who likely included Eratosthenes (see pp. 50–53).

A Eureka Moment

Though little is known for certain about the life of Archimedes, there are some great stories. One such tale, probably the most famous, deals with a crown of gold which was given as a gift to King Hiero II of Sicily (270–215 BCE), who may have been a relative of Archimedes.

Hiero wanted to know if the crown was pure gold or just an alloy. Now, Archimedes would have been able to answer the question quickly if the crown had been a cube or some other regular shape, but unfortunately it was irregular, as crowns tend to be. Knowing that different metals have different densities, that is to say different weights for the same volume, all Archimedes had to do was compare the weight of the crown to the equivalent volume of gold. But the problem that set Archimedes puzzling was: "How do I find the volume of such an irregular shape?"

"Give me a place to stand and I will move the Earth."

—Archimedes

Archimedes imagined himself somewhere in the farthest reaches of space, using a lever to move the Earth to demonstrate the power of simple machines.

Presumably the problem got him all hot and bothered, because Archimedes decided to take the most famous bath in history. As he was getting in he noticed that the water level began to rise. And, reasoning that the water level rose by a volume equivalent to that displaced by his body, Archimedes realized he had found the solution to the problem with the crown.

He was so excited that he ran out into the street yelling "Eureka, Eureka!" Or "I have found it!" We can only guess what his neighbors thought as the genius streaked past.

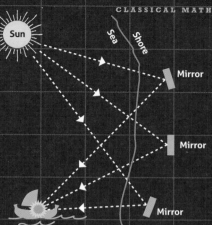

Archimedes' Heat Ray
Reminiscent of little boys with magnifying glasses, Archimedes' heat ray is said to have worked by focusing the power of the Sun to an intense point.

Me and My Death Ray

Another intriguing tale concerns Archimedes' futuristic-sounding "death ray." With Syracuse threatened by Roman invasion, legend has it that Archimedes placed soldiers bearing polished copper shields around the bay in the shape of a parabola (see pp. 102–103). As the fleet approached, the men focused the Sun's reflected rays onto a ship, causing it to burst into flames. Over the years this story has been tested quite a few times, the TV show "MythBusters" being the most recent. Though it's unlikely it worked in this case, it's the same principle that is used to collect TV signals. Your satellite receiver bounces the signals off the dish to collect at the receiver (the node at the end of the arm).

MAJOR WORKS

ON THE EQUILIBRIUM OF PLANES: A two-volume work on centers of gravity, and levers.

ON THE MEASUREMENT OF A CIRCLE: A fraction of a longer work, which contains Archimedes' approximation of π (see pp. 20-21) among other propositions.

ON SPIRALS: Contains a description of the "spiral of Archimedes."

ON SPHERES AND CYLINDERS: Contains the proof that the volume of an inscribed sphere is two-thirds the volume of the cylinder. A sculpted sphere and cylinder were placed on Archimedes' tomb.

ON FLOATING BODIES: Though there is no mention of his "Eureka!" moment in this work, Archimedes gives his principle of buoyancy.

THE SAND RECKONER: Archimedes estimates the number of grains of sand in the universe, and had to create a system for very large numbers to do so.

Archimedean Solids

....

The Platonic solids that we've already met are regular (all of their faces are identical), whereas Archimedean solids are called semi-regular because they have two or more face types.

Stranger Shapes

....

Though the Archimedean solids have different faces, their vertices are regular (in other words they are all the same). These shapes look stranger, and are certainly less familiar in our everyday lives than the Platonic solids. However, looking at the number of faces, edges, and vertices, we can see the relationship the Platonic solids displayed, namely $F + V - E = 2$, also holds true for the Archimedean solids.

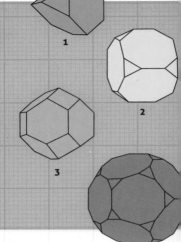

1

2

3

4

5

6

7

8

9

10

11

12

13

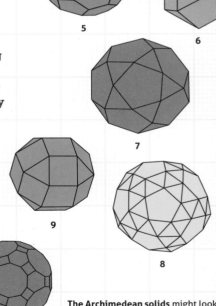

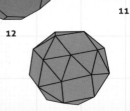

The Archimedean solids might look strange, but one should be a bit more familiar. The fifth shape is essentially the form of some soccer balls—anyone for a game of truncated-icosahedron-ball?

THE ARCHIMEDEAN SOLIDS

Name	Faces	Vertices	Edges
1. Truncated tetrahedron	8 (4 triangles, 4 hexagons)	12	18
2. Truncated cube (hexahedron)	14 (8 triangles, 6 octagons)	24	36
3. Truncated octahedron	14 (6 squares, 8 hexagons)	24	36
4. Truncated dodecahedron	32 (20 triangles, 12 decagons)	60	90
5. Truncated icosahedron	32 (12 pentagons, 20 hexagons)	60	90
6. Cuboctahedron	14 (8 triangles, 6 squares)	12	24
7. Icosidodecahedron	32 (20 triangles, 12 decagons)	30	60
8. Snub dodecahedron	92 (80 triangles, 12 pentagons)	60	150
9. Small rhombicuboctahedron	26 (8 triangles, 18 squares)	24	48
10. Great rhombicosidodecahedron	62 (30 squares, 20 hexagons, 12 decagons)	120	180
11. Small rhombicosidodecahedron	62 (20 triangles, 30 squares, 12 pentagons)	60	120
12. Great rhombicuboctahedron	26 (12 squares, 8 hexagons, 6 octagons)	48	72
13. Snub cube (hexahedron)	38 (32 triangles, 6 squares)	24	60

Eratosthenes

The Moon landing was faked, the 9/11 attacks were an inside job, and the world is flat. It's amazing that the idea of a flat Earth survived for so long in medieval Europe, when so much evidence suggests otherwise. Firstly there's the gradual disappearance of ships on the horizon. Secondly, the Sun, Moon, and to a lesser extent the stars all appear round. Lastly, during an eclipse the shadow cast is round. Given that the scientists of the ancient world had figured this out, it's weird that the idea was later lost.

Life of Eratosthenes

Eratosthenes was a Greek mathematician and astronomer born in Cyrene, which is now part of modern-day Libya, in 276 BCE. Eratosthenes has many achievements to his name. As an astronomer, he calculated the circumference of the Earth, the distance from the Earth to the Moon, and the distance from the Earth to the Sun. As a student he studied under Ariston of Chios, a student of Zeno. As a teacher he tutored Philopater, the son of Ptolemy III, whose grandfather was Ptolemy I, one of Alexander the Great's generals.

As a mathematician he is known for a method of determining prime numbers called "the sieve of Eratosthenes." As a geographer he made the first maps of the known

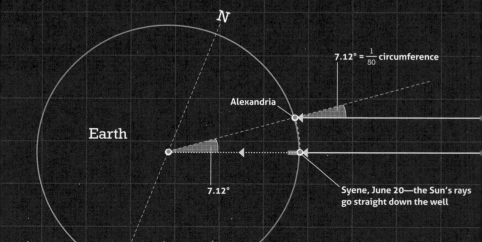

$7.12° = \frac{1}{50}$ circumference

N

Alexandria

Earth

7.12°

Syene, June 20—the Sun's rays go straight down the well

world and charted the Nile.
Eratosthenes also gave us a
calendar with a leap year. Given
these accomplishments it's perhaps
a little unkind that Eratosthenes was
given the nickname "beta" or
"second" because he was good at
so many things, but never the best.
In his last years Eratosthenes went
blind and it's said that in 195 BCE he
committed suicide by starvation.

Around the World

Eratosthenes' calculation of the
circumference of the Earth is
incredible (see below), though
some doubt its validity given the
many variables. Nevertheless,
it's a great exercise in math.

Eratosthenes noticed that at noon
during the summer solstice there
were no shadows in the town of
Syene, now Aswan, Egypt. (Actually
there would have been a small
shadow because Aswan is slightly
north of the Tropic of Cancer.)

> Eratosthenes noticed that
> at noon during the summer
> solstice there were no
> shadows in the town of
> Syene, now Aswan, Egypt.

He then measured the angle of the
shadow during the summer solstice
in Alexandria to be 7° 12' or 7.2°—
a very accurate measurement.
This represented $\frac{72°}{360°}$ or $\frac{1}{50}$ of the
circumference of the Earth. That
was the easy part.

So now he knew that the distance
from Syene to Alexandria is $\frac{1}{50}$ of
the Earth's circumference, but how
far is it from Syene to Alexandria?
Eratosthenes stated that it was 5000
stadia, but how far is a stadion?
Various estimates peg the value

Eratosthenes' ingenious way of working out
the circumference of the Earth, based on the
shadows that fell on the summer solstice.

Distance between markers = 5000 stadia

Sun

SIEVE OF ERATOSTHENES

Eratosthenes' eponymous method for identifying prime numbers is simple and effective. It's good for the smaller primes, less so for the larger ones, largely due to the amount of space required. In the example below, we find the primes lower than one.

Using a ten-by-ten grid, with the squares numbered from one to a hundred, the first step is to highlight the first prime, two, and then mark off all its multiples. (Given that these numbers are divisible by two, they are of course not primes.) After this, you simply need to repeat the trick for the next three primes: three, five, and seven. You'll see as you progress that some of the multiples have already been marked off (for example, seven threes is the same as three sevens). By the time you reach the fifth prime, eleven, only the prime numbers will remain unmarked.

	2	3	4	5	6	7	8	9	10
11	12	13	14	15	16	17	18	19	20
21	22	23	24	25	26	27	28	29	30
31	32	33	34	35	36	37	38	39	40
41	42	43	44	45	46	47	48	49	50
51	52	53	54	55	56	57	58	59	60
61	62	63	64	65	66	67	68	69	70
71	72	73	74	75	76	77	78	79	80
81	82	83	84	85	86	87	88	89	90
91	92	93	94	95	96	97	98	99	100

Initial grid
In this example, we are using a ten-by-ten grid to identify the primes lower than one hundred.

	2	3	4	5	6	7	8	9	10
11	12	13	14	15	16	17	18	19	20
21	22	23	24	25	26	27	28	29	30
31	32	33	34	35	36	37	38	39	40
41	42	43	44	45	46	47	48	49	50
51	52	53	54	55	56	57	58	59	60
61	62	63	64	65	66	67	68	69	70
71	72	73	74	75	76	77	78	79	80
81	82	83	84	85	86	87	88	89	90
91	92	93	94	95	96	97	98	99	100

The first prime
The first prime is highlighted in green, its multiples in orange.

	2	3	4	5	6	7	8	9	10
11	12	13	14	15	16	17	18	19	20
21	22	23	24	25	26	27	28	29	30
31	32	33	34	35	36	37	38	39	40
41	42	43	44	45	46	47	48	49	50
51	52	53	54	55	56	57	58	59	60
61	62	63	64	65	66	67	68	69	70
71	72	73	74	75	76	77	78	79	80
81	82	83	84	85	86	87	88	89	90
91	92	93	94	95	96	97	98	99	100

Next up is three
The multiples of three are marked in orange too.

	2	3	4	5	6	7	8	9	10
11	12	13	14	15	16	17	18	19	20
21	22	23	24	25	26	27	28	29	30
31	32	33	34	35	36	37	38	39	40
41	42	43	44	45	46	47	48	49	50
51	52	53	54	55	56	57	58	59	60
61	62	63	64	65	66	67	68	69	70
71	72	73	74	75	76	77	78	79	80
81	82	83	84	85	86	87	88	89	90
91	92	93	94	95	96	97	98	99	100

Take five
Now is the turn of the multiples of five.

PLATONICUS (LOST):
A work that dealt with the mathematics of Plato's philosophy. Although it's now lost, we know of it via the Theon of Smyrna (ca. 70-135 CE), who worked with prime and geometrical numbers as well as music.

ON MEANS (LOST):
Another lost work on geometry. Mentioned by the Greek geometer Pappus (290-350 CE), who claimed it to be one of the greatest books on geometry.

ON THE MEASUREMENT OF THE EARTH (LOST):
In this work, which is also lost, Eratosthenes calculated the circumference of the Earth. We know of it via Greek astronomer Cleomedes (ca. 10-70 CE) and Theon of Smyrna.

· ·

somewhere between 157 m and 185 m. The distance from Alexandria to Aswan is 843 km; dividing that by 5000 would make a stadion 169 m, assuming the measurer moved in a direct line from one place to the other. However, this simple assumption is actually quite unlikely because Eratosthenes stated that Alexandria is to the north of Syene when in fact it's to the northwest. Also, you have to consider that traveling in a straight line without any of our modern gadgets would be pretty difficult, what with the desert and a twisting Nile in your way.

So let's accept all the possible errors—there are more besides— and calculate the upper and lower values for the Earth's circumference. If we take 5,000 stadia to be $\frac{1}{50}$ of the circumference then 250,000 stadia would be the total circumference. At 157 m per stadion that makes the circumference of the Earth 39,250 km; at 185 m per stadion it's 46,250 km. With the true circumference a a fraction over 40,000 km, the first result is approximately a couple of percent low, while the second is 16 percent high. All in all, a remarkably accurate estimate.

	2	3	4	5	6	7	8	9	10
11	12	13	14	15	16	17	18	19	20
21	22	23	24	25	26	27	28	29	30
31	32	33	34	35	36	37	38	39	40
41	42	43	44	45	46	47	48	49	50
51	52	53	54	55	56	57	58	59	60
61	62	63	64	65	66	67	68	69	70
71	72	73	74	75	76	77	78	79	80
81	82	83	84	85	86	87	88	89	90
91	92	93	94	95	96	97	98	99	100

	2	3	4	5	6	7	8	9	10
11	12	13	14	15	16	17	18	19	20
21	22	23	24	25	26	27	28	29	30
31	32	33	34	35	36	37	38	39	40
41	42	43	44	45	46	47	48	49	50
51	52	53	54	55	56	57	58	59	60
61	62	63	64	65	66	67	68	69	70
71	72	73	74	75	76	77	78	79	80
81	82	83	84	85	86	87	88	89	90
91	92	93	94	95	96	97	98	99	100

	2	3		5		7			
11		13				17		19	
		23						29	
31						37			
41		43				47			
		53						59	
61						67			
71		73						79	
		83						89	
						97			

Seven's up
The first multiple of seven yet to be highlighted orange is forty-nine. There are eight in all to be marked...

Any others?
The remaining unmarked numbers are not divisible by any of the preceding numbers...

Final result
...and nor do they multiply in any of the unmarked numbers that follow them. All of them are therefore prime...

Diophantus

Diophantus is yet another enigma of the ancient world. Some mathematicians call him the "father of algebra" for his work *Arithmetica*, while others give that title to Al-Khwarizmi (see pp. 70–71). Diophantus's claim to the title is based on the fact that it was in his works that the transformation from language-based mathematics to the symbol-based mathematics that we know and work with today occurred.

The Riddle of Diophantus

For a mathematician whose influences are still felt today, very little is known about Diophantus of Alexandria. Just pinning down when he lived is tough enough, and what is known about him comes mostly from secondary sources and what remains of his writing.

Diophantus lived in Alexandria during the third century. Most guess that his birth was around 200 CE and his death some eighty-four years later. This estimate come from Diophantus quoting Hypsicles (190–120 BCE), a Greek mathematician who worked with regular polyhedra, which places him some time after 150 BCE; while Theon of Alexandria (335–405 CE), another Greek mathematician and the father of Hypatia—the first known female mathematician—quotes Diophantus, which places his death before 350 CE.

His eighty-four-year lifespan comes from "The Riddle of Diophantus," which was taken from a fifth-century Greek anthology of number games. Many slightly different English versions exist, but the following is quite nice:

MAJOR WORKS

ARITHMETICA: Arithmetica is a collection of problems (some say there are 130, others say 189) giving numerical solutions to determinate (a defined limit to the solution in one variable) and indeterminate equations (an infinite number of solutions in two or more variables).

THE ARITHMETICA: Consisted of thirteen books, six of which survive. There are also four Arabic books that some think are translations of Diophantus's work. The books solve problems involving linear and quadratic equations, but Diophantus considered only positive rational solutions—in other words he ignored zero and negative numbers. The books were translated into Latin by Bombelli in 1570, and influence European mathematics through to the modern day. In fact, a 1621 translation by Claude Bachet, a French mathematician who wrote books on mathematical puzzles,

The riddle has Diophantus living eighty-four years, but as you can imagine a word problem is not the most authoritative source. The riddle is worth solving nonetheless. If we assign x as his age at the time of his death, then the equation to find x would be:

$$x = \frac{x}{6} + \frac{x}{12} + \frac{x}{7} + 5 + \frac{x}{2} + 4$$

"Here lies Diophantus,
   the wonder behold

Through art Algebra,
   the stone tells how old:

God gave him his boyhood
   one-sixth of his life,

One-twelfth more as youth
   while whiskers grew rife;

First we collect our xs to one side:

$$x - \frac{x}{6} - \frac{x}{12} - \frac{x}{7} - \frac{x}{2} = 9$$

And then yet one-seventh
   ere marriage begun;

Then we get a common denominator for the fractions, it's 84, and we rewrite the equation:

In five years there came
   a bouncing new son.

$$\frac{84x}{84} - \frac{14x}{84} - \frac{7x}{84} - \frac{12x}{84} - \frac{42x}{84} = 9$$

Alas, the dear child
   of master and sage

Now we subtract the numerators to get:

After attaining half the
   measure his father's life
   chill fate took him.

$$\frac{9x}{84} = 9$$

After consoling his fate by
   the science of numbers
   for four years, he ended
   his life."

Next we multiply both sides by 84 then divide by 9 to get:

$$x = 84$$

caused Pierre de Fermat (ca. 1601-65) to write in the margin 'I have discovered a truly wonderful proof, but the margin is too small to contain it.' It took over three hundred years for mathematicians to solve 'Fermat's Last Theorem.'

PORISMS (LOST):
In *Arithmetica*, Diophantus makes reference to another work, *Porisms*, which is completely lost, although fractions of another work, *On Polygonal Numbers*, still exist.

3.

An
International
Language

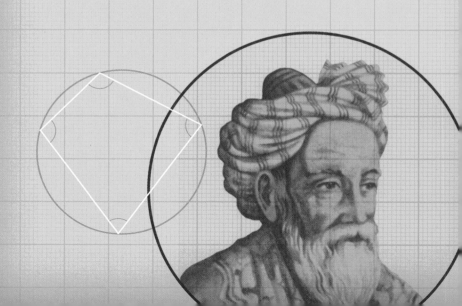

As we mentioned at the start of the previous chapter, our familiarity with the ancient Greeks perhaps blinds us to the mathematical advances made elsewhere. However, the fact is that key figures such as Brahmagupta, Al-Khwarizmi, and Omar Khayyam, alongside many other Eastern mathematicians, made contributions to our modern understanding of math that were arguably even more profound than those of the ancient Greeks.

Egyptian Mathematics

Ancient Egyptian mathematics was really quite advanced. Like us, the ancient Egyptians had a base-ten system, but unlike ours it wasn't a place-value system. Instead they had separate symbols to represent the one, ten, a hundred, a thousand, ten thousand, and a million. I love the kneeling guy for a million; I can just picture some ancient Egyptian on his knees shouting: "Yes, I've won a million... I'm rich!"

Math in the Time of the Pharaohs

The Egyptian hieroglyphs used to represent numbers are shown below; and it's easy to use these to represent other numbers. For example, one is represented by the figure ı, so any number up to nine can be represented by that many ı; for instance, three would be ııı.

You can also make larger numbers quite simply. When you reach the next hieroglyph up, you simply use that at the beginning; for example, 123 would be written as ℮⋒ıı.

Adding and subtracting is also easy, and uses a system of carrying forward and borrowing, much like we do today. As an example, let's add two numbers: twenty-eight and 103; or in Egyptian numerals ⋒⋒ıııııııı (28) and ℮ııı (103).

We just add the units together to get eleven, which is a one and a ten. That is to say ıııııııııııı = ⋒ı, which is like the carrying forward we do in our base-ten arithmetic. Then we do the same for the tens and the hundreds. So:

28 + 103 = 131 or ⋒⋒ıııııııı + ℮ııı = ℮⋒⋒⋒ı

Egyptian numeral hieroglyphics It's interesting that the Egyptians used a base-ten number system, while later the Greeks and Romans did not. The Greeks and Romans included symbols for other values such as five, fifty, and so on. The base-ten system would only reappear from the East many centuries later.

1	10	100	1000	10,000	100,000	1,000,000

Multiplication and division are a little more complicated, but the method used is quite interesting. What the Egyptians did was to successively double one of the two numbers. Let's try 11 • 26.

∩∩ıııııı = one 26

∩∩∩∩∩ıı = two 26s

ℓıııı = four 26s

ℓℓıııııııı = eight 26s

Now to make eleven 26s they would just add the eight, two, and one 26s that we worked out above. We have a total of sixteen units so we carry the ten to the next symbol and leave six units. We now have a total of eight tens, and two hundreds, which can be written as:

ℓℓ∩∩∩∩∩∩∩∩ııııı = 286

Ancient Egyptian Fractions
....

The ancient Egyptians also had a method for dealing with fractions. To write a fraction they would just add an "eye" symbol above the number that represented the divisor; so $\frac{1}{2}$ would look like ⌢ı̈ı and $\frac{1}{10}$ would be ⌢∩.

This limited the Egyptians to using only unit fractions—those fractions where the numerator (the top number) is always one. But to make other fractions they would just add unit fractions together; for example, $\frac{5}{6}$ would be $\frac{1}{2} + \frac{1}{3}$ or ⌢ı̈ı + ⌢ı̈ıı.

3000-YEAR-OLD HOMEWORK

The Rhind Papyrus is named after the Scottish Egyptologist, A. Henry Rhind who purchased the papyrus in Egypt in 1858. It's a scroll about 20 ft (6 m) long and 1 ft (30 cm) wide. It was written around 1650 BCE by Ahmes, a scribe who states that he was copying an earlier text that was a couple of hundred years older still. Therefore the material on the papyrus possibly dates from as long ago as 1850 BCE.

Some eighty-seven problems are contained on the papyrus, ranging from those dealing with basic arithmetic—though with Egyptian numerals, division and multiplication were not that easy—to geometry and equation solving. Though the papyrus deals with problems that would involve equations, they are not like those we know—the beginnings of algebra as we understand it would not appear for centuries.

The Moscow Papyrus, or the Golenishchev Papyrus, is roughly 15 ft (4.5 m) and 3 in (7 cm) wide. It consists of twenty-five problems that are mostly geometric in nature.

Completing the Square

As we've seen, quadratics equations (where x^2 is the leading term) have been studied for thousands of years, and there are a number of ways to solve them. As a taster to the topic, we'll take a second to work through one such problem using a nicely visual technique called "completing the square."

Balooning Around

The problem goes like this: A child fires a balloon over a fence and it follows a parabola (see pp. 102–103), which can be represented by $h = -x^2 - 6x + 40$, where h is the height and x is the distance (in meters) to the left and right of the fence. How far behind the fence was the kid when he shot the balloon and how far over the fence did it land?

The first thing we should note is that we're trying to find out when the balloon hits the ground. The ground represents a height of zero, therefore $h = 0$. The equation will now be:

$$0 = -x^2 - 6x + 40$$

For simplicity, we will add x^2 and $6x$ to both sides to get the equation:

$$x^2 + 6x = 40$$

This helps us because the positive numbers allow a geometric solution. Now x^2 can represent a square with sides that are both of the length x, and $6x$ can represent a rectangle with one side that is x long and one that is 6 long.

Adding these two shapes, as the left-hand side of the equation ($x^2 + 6x$ …) indicates, makes a rectangle that measures x on one side and $x + 6$ on the other, in other words $x \cdot (x + 6)$.

At this point we could guess what x is, and if the number happens to be an integer we might get lucky. But if the answer is rational or irrational then things get trickier.

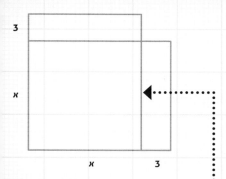

Diophantus called negative answers absurd, but in theoretical mathematics they are answers nonetheless.

Being Square

We're "completing the square," not a rectangle, so we need to divide the $6x$ rectangle into two equal rectangles, both measuring $3x$, and place them on either side of the square.

Now we complete the square by filling in the top right corner. The little square measures 3 by 3, so we perform the simple multiplication to find that we need to add 9—remember to add it to both sides. This gives:

$$(x + 3)^2 = 49$$

We now have to ask ourselves, what number, when squared, is equal to 49? Well, 7^2 is 49, therefore $x + 3$ must equal 7 and that would mean $x = 4$.

This makes a lot of sense, but there is another value of x that is not at all obvious from this geometric approach.

Diophantus (see pp. 54–55) called negative answers absurd, but in theoretical mathematics they are answers nonetheless. These must be considered when solving the problem. In this case, $(-7)^2$ also equals 49, therefore $x + 3$ must equal -7 as well, and that would make $x = -10$. Let's face it, it's pretty hard to draw a square that measures -10, thus the aversion to negative numbers when you view the problem from a geometric stance. Taking these values of -10 and 4 means the child0 was 10 meters to the left of the fence when he shot the balloon.

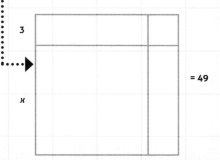

Indian Mathematics

We all use numbers in our everyday lives, but most of us do it without ever questioning where they're from. We're far more familiar with the origins of words and the development of our language; but numbers play a crucial role in our lives, so why do we never ask: "Where do our numbers come from?"

The Indian Sulbasutras

Sulbasutras are appendices to religious texts. They were not theoretical texts, but rather applied mathematics relating to the construction of religious architecture.

In the sulbasutra of Baudhayana (ca. 800–740 BCE) the Pythagorean theorem—or at least a special case of it, the right-angled isosceles triangle—is dealt with. Again, our reference to Pythagoras, given that he would not be born for another two hundred years, reveals our Eurocentric view of the world. In the sulbasutra of Katyayana (ca. 200–140 BCE) there is reference to the theorem without restrictions, though this was after the time of Pythagoras.

Somewhat earlier, two sulbasutras, that of Apastamba, who lived in the sixth century BCE, and Katyayana (third century BCE), gave a value for the square root of two of $\frac{577}{408}$, which is accurate to the fifth decimal place.

Given that these texts deal with construction, circles come into play. Interestingly, the approximations for the value of π change depending on the nature of the calculation. It seems the applied nature of the sulbasutras made an exact value unnecessary. Values used for π ranged from about 3 to 3.2.

The Indian Numerals

The true significance of the Indian contribution was commented upon by the French mathematician Pierre-Simon Laplace (1749–1827):

"It is India that gave us the ingenious method of expressing all numbers by means of ten symbols, each symbol receiving a value of position as well as an absolute value; a profound and important idea which appears so simple to us now that we ignore its true merit. But its very simplicity and the great ease which it has lent to computations put our arithmetic in the first rank of useful inventions; and we shall appreciate the grandeur of the achievement the more when we remember that it escaped the genius of Archimedes and Apollonius, two of the greatest men produced by antiquity." (Quoted in H. Eves, *Return to Mathematical Circles*; 1988.)

Laplace is right; imagine if we still tried to work with Roman numerals. The advancements we have made would certainly have been delayed. We just have to review the difficulty in multiplying Egyptian numbers, which were at least part of a base-ten system, Roman numerals would have been worse.

Brahmi numerals (see p. 67), which came into use around 250 BCE, were part of a kind of base-ten system. It had distinct symbols for the first nine numbers and separate symbols for multiples of ten and one hundred. That is to say there was a symbol for twenty, thirty, four hundred, five hundred, and so on. In the seventh century CE, around the time of Brahmagupta (see pp. 64–67), a base-ten positional system came into use. It's interesting that the Egyptians had a base-ten system, but not a place-value one while the Babylonians had a positional system that was base-sixty rather than ten.

Aryabhatiya
....

Aryabhata wrote the *Aryabhatiya*, which collected all the mathematics in India up to that point—very much like Euclid collecting the rules of geometry (see pp. 44–45). The *Aryabhatiya* contains arithmetic, algebra, trigonometry, and quadratic equations, and also a very accurate value for π (3.1416). The difference, though, was that Euclid provided rigorous proofs for the rules. Then, in the sixth century, Varahamihira summarized astronomical works as well as looking at Pascal's triangle (see pp. 116–21) and magic squares (see box below).

Next came Brahmagupta who lived in the seventh century, and whose work was extended by Mahavira in the ninth century. A generation on, and Prthudakasvami continued the work on the algebra of quadratic equations; while Sridhara (870–930 CE) was one of the first to produce a general formula for solving quadratics, though for only one root.

MAGIC SQUARES

A magic square is a square where the rows, columns, and diagonals all add to the same amount. In the cases shown here that is thirty-four and fifteen.

13	8	10	3
12	1	15	9
7	14	4	6
2	11	5	19

8	1	6
3	5	7
4	9	2

Brahmagupta

Sometimes the littlest things are the most important and simultaneously the most overlooked. For most people, this little thing is so small it's nothing, quite literally zero. It was Brahmagupta who gave us a new way of looking at zero. Mathematics up to this time had been hampered by numeral systems that would have made the calculations we do today far more difficult. The ancient Egyptians had a base-ten system, and the Babylonians had a place-value system, but it was Indians who gave us the base-ten place-value system we know today. Moreover, Brahmagupta was also the first to investigate zero as a number, not just a place holder.

Star Potential

Brahmagupta was born in 598 CE in the city of Bhinmal in the northwest of India near today's border with Pakistan. He was made head of the observatory in Ujjain, a city east of Bhinmal and a center of astronomy and mathematics. During his time in Ujjain, Brahmagupta wrote several texts. The *Brahmasphutasiddhanta* is the most famous of his works.

PELL'S EQUATIONS

Pell's equations are in the form $x^2 - ny^2 = 1$. They are named after John Pell, an English mathematician (1611-85) who really had little to do with their development (similar work had been done by Diophantus). Brahmagupta was the first to truly study Pell's equations.

What makes these equations interesting is that the integer solutions approximate the root of n. That is to say the solutions to $x^2 - 2y^2 = (1)$ (3, 2), (17, 12), (577, 408) and so on, are better and better approximations for the square root of two.

$$\frac{3}{2} = 1.5$$

$$\frac{17}{12} = 1.416$$

$$\frac{577}{408} = 1.414215686$$

where $\sqrt{2} = 1.414213563$

"0"

He also produced the *Cadamekela*, *Khandakhadyak*, and *Durkeamynarda* but very little information can be found on these works today. He died in 670 CE.

Brahmasphutasiddhanta

In 628, Brahmagupta wrote the *Brahmasphutasiddhanta*—now that's a tongue twister—a text which has had a profound affect on Western mathematics.

The *Brahmasphutasiddhanta* is divided into twenty-five chapters. The first ten are thought to be an earlier work of Brahmagupta's, while the next fifteen are improvements or addendums to the first ten. In the *Brahmasphutasiddhanta* several different topics are covered. Diophantine analysis (see pp. 54–55) is covered in chapter twelve, where Brahmagupta looks at Pythagorean

LAYING DOWN THE LAW

The significance of Brahmagupta's laws is that they treat zero as a number, not just a place holder, and negative numbers are treated as numbers, not numerical outcasts to be ignored.

1) A zero added to a number is the number.

2) A zero subtracted from a number is the number.

3) A number times zero is zero.

4) A negative minus zero is a negative.

5) A positive minus zero is a positive.

6) Zero minus zero is zero.

7) A negative subtracted from zero is a positive.

8) A positive subtracted from zero is a negative.

9) The product (what you get when you multiply numbers) of zero multiplied by a negative or positive is zero.

10) The product of zero multiplied by zero is zero.

11) The product or quotient (what you get when you divide numbers) of two positives is one positive.

12) The product or quotient of two negatives is one positive.

13) The product or quotient of a negative and a positive is a negative.

14) The product or quotient of a positive and a negative is a negative.

triples (see p. 37) and a family of equations now known as Pell's equations (see box on p. 64). Brahmagupta also developed an equation for the area of the "cyclic quadrilateral" (a four-sided shape where each point touches the inside of a circle—see below).

$$a + c = 180°$$
$$b + d = 180°$$

Cyclic quadrilaterals are four-sided shapes where the vertices (corners) all touch a circle. They have neat properties, including that their opposite internal angles add to 180°.

However, the most important section of the *Brahmasphutasiddhanta* deals with zero and negative numbers, this covers the rules that we learn for integers around the age of twelve (see box on p. 65).

The *Brahmasphutasiddhanta* treated zero and negative numbers as potential solutions. Previously, with a geometric approach, zero and negatives had been ignored or considered absurd because in the real world zero or negative lengths or areas simply don't exist. However, as we know today, these numbers have many practical applications— I might not be able to hold a negative dollar in my hand, but I can sure as hell see them in my bank account.

Brahmagupta hit a few snags along the way, especially when it came to dividing by zero. But this gives most people headaches; just ask anyone who uses calculus, whether in science, engineering, business, or medicine.

Previously, with a geometric approach, zero and negatives had been ignored or considered absurd because in the real world zero or negative lengths or areas simply don't exist.

Indian–Arabic Numerals

The numerals we use today have a evolved over thousands of years, from Indian origins in the first century CE through Arabic variations around one thousand years later. The numeric forms we're familiar with today emerged approximately six hundred years ago.

1st century	5th century	10th century	11th century	12th century	14th century
—	৩	৭	I	٦	٦
≡	ᾧ	₹	८	٢	٢
≣	ᾧ	₹	₹	₹	3
⊁	ᾨ	✗	✗	৪	4
Ⱶ	Ⱶ	५	५	९	5
6	ৎ	ऽ	১	৫	৫
)		ৗ	ৗ	ৗ	7
ৎ		ৎ	৪	৪	৪
২		৶	৭	৭	৭
	ৗ	৹		ᴓ	৹

Math in the Middle East

I remember learning about the Egyptians, Greeks, and Romans one year in school. Then at the start of the next year there was a cursory mention of the "Dark Ages," before boom ... we were in the middle of the Renaissance. It was like we went from the fall of the Western Roman Empire (late fifth century CE) to the dawning of the Renaissance almost a millennium later, with nothing in between. As it turns out, there was quite a lot going on; during this time the center of learning shifted east to Baghdad.

The House of Wisdom

The closing of Plato's academy in 529 CE could be considered Greek mathematics' last gasp. From that point until the thirteenth century the mathematical center of the world was in the East.

The House of Wisdom in Baghdad was established by Harun Al-Rashid (763–809), the fifth Caliph of the Abbasid dynasty, and his son Al-Ma'mun (786–833). During this time the Islamic empire stretched from Spain in the west to the borders of India in the east. Originally the House of Wisdom was focused on translating and preserving works from Persia, then Greece, and India. It became a center for the humanities and sciences, but was razed during the Mongol invasion of 1258.

Though little is said about Persian mathematicians, there have been many of significance. Starting around the time of the founding of the House of Wisdom, we have Al-Khwarizmi (see pp. 70–71). Then Al-Kindi, who lived from 801 to 873, and wrote on Indian number systems. Around the same time, the three Banu Musa brothers worked on geometry, astronomy, and mechanics.

Abu Kamil, who was born in 850 and died in 930, extended Al-Khwarizmi's work on algebra. Then Ibrahim ibn Sinan, who was born in 908 and lived for only thirty-eight years, extended the theory of

Arabic numerals began as Indian numerals. The earliest surviving text that contains them is from the tenth century.

integration, taking Archimedes' method of exhaustion further. Finally, Al-Karaji (953–1029) made significant advances in algebra, making it much more like what we know today and less like geometry. In addition, there are a host of other mathematicians who either translated texts, made commentaries on text, or furthered mathematics in the areas of geometry, trigonometry, number theory, and so on.

Arabic Numerals

One of the most significant advances made during this time was the adoption of a base-ten place-value system for numbers. In other words, a system that has ten symbols, and within which the place of the symbol determines its value. Sound familiar? Well, the system we work with today is a base-ten place-value system. That is to say when we write the number 535, we know the numerals have different values based on their position: the first five are "hundreds" and the last five "units."

As we've already seen, the ancient Egyptians had a base-ten system, but not a place-value one—they had a different symbol for each power of ten. This made things like multiplication very difficult. Roman numerals were even more difficult to deal with. The base-ten place-value system, however, allows for easier methods of calculation and the development of decimals. This matters because making arithmetic simpler frees up mathematicians to think about the bigger picture—consider how much easier the number-crunching power of computers makes it for today's scientists to think profound thoughts.

Arabic numerals began as Indian numerals. A Christian bishop living near the Euphrates River wrote of their use in 662, but the earliest surviving text that contains them is from the tenth century. A twelfth-century Latin text *Algoritmi de Numero Indorum*, (a translation of an Al-Khwarizmi text), is often claimed to be the first Arabic text on Indian numerals. This dates the adoption of Indian numerals to between 790 and 840. It's interesting to note that our word "algorithm" comes from the title of this text. In 1202 Fibonacci introduced Arabic numerals to Europe.

By the time of this fourteenth-century treatise, Arab mathematicians had played a central role in the evolution of math.

Al-Khwarizmi

Much like Euclid, the Greek "father of geometry" and the author of one of the greatest books of all time, *Elements*, very little is known about Al-Khwarizmi. Most people will find this amazing, as they've never heard of him, but it's from Al-Khwarizmi that we get two household words—at least they are if you live in my house—algebra and algorithm.

Polymath

Al-Khwarizmi, or Abu Ja'far Muhammad ibn Musa Al-Khwarizmi, was born around 780 CE, although exactly where he was born is a subject of some debate. Some think that he came from Khwarezm, which is now in Uzbekistan, south of the Aral Sea and east of the Caspian, while others suggest he was born in Baghdad.

"Al-Jabr"

One thing that is known for sure is that Al-Khwarizmi worked at the House of Wisdom (see pp. 68–69). When there, he worked with the Banu Musa brothers as translators of Greek, Indian, and other texts. He also attempted to expand on these translations by writing his own texts on algebra, geometry, astronomy, and geography.

Al-Khwarizmi completed several important texts. One major work was *Kitab Surat Al-Ard*, written in 833. It revised Ptolemy's *Geography*, including the coordinates of over 2,400 cities and geographical features. In *Kitab Surat Al-Ard*, Al-Khwarizmi corrected Ptolemy's overestimation of the length of the Mediterranean Sea and added details about lands to the east, which were better known to the Abbasid dynasty than to the Greeks. Al-Khwarizmi also wrote several minor works on astrolabes (an instrument used by astronomers, astrologers, and navigators), sundials, and the Jewish calendar.

The Origins of "Algorithm"

Al-Khwarizmi's second most important work was *Algoritmi de Numero Indorum*, the title of a Latin translation of his original Arabic text, which has been lost. It's from this title that we get the word "algorithm," which means a number of steps or instructions to be followed. In this work Al-Khwarizmi introduces the Hindu place-value system for numerals. It's also believed that this is the first use of zero as a place holder. He also gives methods for calculations and also a method for finding square roots.

The Origins of "Algebra"

Hisab Al-Jabr w'Al-Muqabala is Al-Khwarizmi's most significant work, and it's from the "Al-Jabr" part of the title that we get the word algebra. Though some scholars consider Diophantus (see pp. 54–55) to be the "father of algebra," others believe that title belongs to Al-Khwarizmi because of his work in this text.

Al-Khwarizmi's method for solving linear and quadratic equations was to reduce the equations to one of six forms. This meant that Al-Khwarizmi could neatly sidestep the problems that were posed by negative numbers. At this point it's interesting to define the parts of a quadratic according to Al-Khwarizmi. Given $ax^2 + bx + c = 0$ where a, b, and c are numbers, ax^2 represents the square, bx represents the root, and c represents the number. The six forms Al-Khwarizmi allowed were:

1) Squares equal to roots or $ax^2 = bx$

This would be an equation like $x^2 = 4x$ which gives an answer of 4; and $3x^2 = 7x$ which gives an answer of $\frac{7}{3}$. These solutions are quite elementary when studied more closely. If we take

the second example and divide both sides by 3 we get $x^2 = \frac{7}{3}x$ and since x^2 or $x \cdot x$ is on the left and $\frac{7}{3}x$ or $\frac{7}{3} \cdot x$ is on the right then:

$$x \cdot x = \frac{7}{3} \cdot x$$

and the first x on the left must be $\frac{7}{3}$. It's interesting to note that the first obvious solution for x is not given, namely $x = 0$.

2) Squares equal to numbers or $ax^2 = c$

Though I can't find any reference to show Al-Khwarizmi's method, a simple approach would be to isolate the x by dividing both sides by a and determining the square root by some method, possibly something similar to Archimedes' method.

3) Roots equal to numbers or $bx = c$

4) Squares and roots equal numbers or

$$ax^2 + bx = c$$

5) Squares and numbers equal roots or

$$ax^2 + c = bx$$

6) Roots and numbers equal squares or

$$bx + c = ax^2$$

According to American historian of mathematics, Carl Boyer (1906–76), writing in *A History of Mathematics* (1968), for these three last examples: "The solutions are 'cookbook' rules for 'completing the square' applied to specific instances." We've already solved quadratics by completing the square on pp. 60–61.

Omar Khayyam

Omar Khayyam was the first non-European mathematician I became aware of. From this point my Eurocentric ideas started to change, though from the contents of this book I'm sure you can see there is still room for improvement.

Math Prodigy

Omar Khayyam was born on May 18, 1048, in Nishapur, Persia (now Iran). Before the age of twenty-five, Khayyam had already produced significant works in mathematics. In 1070, he moved to Samarkand, in modern-day Uzbekistan, where he was supported by Abu Tahir, a prominent jurist. This allowed Khayyam to write his most important work: *Treatise on Demonstration of Problems of Algebra*. In 1073 Malik-Shah, sultan of the Seljuk dynasty, invited Khayyam to the city of Esfahan, the capital, to set up an observatory. Khayyam remained in Esfahan for the next 18 years. In 1092, after the death of Malik-Shah, political unrest ensued until 1118 when Malik-Shah's third son Sanjar assumed control of the Seljuk Dynasty. Sanjar moved the dynasty's capital to Merv, and Khayyam moved there sometime after 1118. In Merv, another center for learning was created, and Khayyam continued to work on his mathematics, until he died on December 4, 1122.

 MAJOR WORKS

THE RUBAIYAT:
Khayyam is probably best known through the *Rubaiyat of Omar Khayyam*, a translation by Edward Fitzgerald. *The Rubaiyat* is a collection of 600 quatrains (four-line poems).

PROBLEMS OF ARITHMETIC:
A book on algebra and music. At the request of Malik-Shah, Khayyam set up an observatory in Esfahan. He calculated the length of a year to be 365.24219858156 days, which is amazingly accurate. He also created a new calendar, the Jalali calendar.

COMMENTARIES ON THE DIFFICULT POSTULATE OF EUCLID'S BOOK:
Analyzing Euclid's parallel postulate, Khayyam made inroads into non-Euclidean geometry, although some claim that these were unintentional.

PROBLEMS OF ALGEBRA (LOST):
In his other works, Khayyam makes reference to a lost work in which he writes about what would later become known as Pascal's triangle (see pp. 116-21).

"The majority of people who imitate philosophers confuse the true with the false, and they do nothing but deceive and pretend knowledge, and they do not use what they know of the sciences except for base and material purposes.**"**

—*Treatise on Demonstration of Problems of Algebra*

The Problems of Algebra

Treatise on Demonstration of Problems of Algebra is Omar Khayyam's greatest mathematical work. In this book, written in 1070, Khayyam outlines a complete classification of cubic equations with solutions found by using conics (see box on the next page).

By finding the intersection point of two conics by geometric means, Khayyam was able to solve cubic equations. However, it's interesting to note that he only found one or maybe two of the three possible solutions.

His solutions were geometric in nature, but Khayyam hoped that one day an arithmetical solution would be developed. This occurred many centuries later in the works of Italian mathematicians.

Omar Khayyam used geometry, and shapes such as the parabola and circle, to solve cubic equations. He argued that, despite appearances to the contrary, geometry could be used to solve algebraic problems and pointed to Euclid's *Elements* as proof of that fact.

Conics

Conics is a section of mathematics that deals with shapes that can be drawn from a cone, or more accurately from two cones placed nose to nose. There are four shapes that can be created from a cone—the circle, the ellipse, the parabola, and the hyperbola—and each can be used to solve specific algebra problems. They are an important part of the history of mathematics and have been studied for thousands of years.

The Four Shapes

As is shown on the next page, these four conics appear in the world in a wide range of contexts, both man-made and natural.

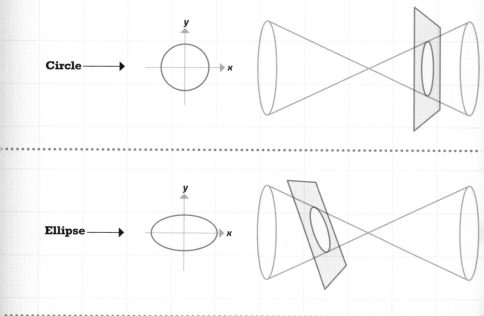

Circle ⟶

Ellipse ⟶

Conics Are Everywhere

These shapes show up all the time, in contexts that will be familiar to many people, even though they may not be aware that what they are seeing is a conic! The circle is an obvious case; it's pretty tough to have an automotive industry without circular tires. Meanwhile, an example of the ellipse is the path the Earth follows around the Sun; an example of a parabola (spun in three dimensions) is a satellite dish; while the shadow cast by a lampshade on a wall is a hyperbola.

Intersecting Lines

An additional branch of conics, known as degenerate conics, is simply the intersection of two lines at a single point.

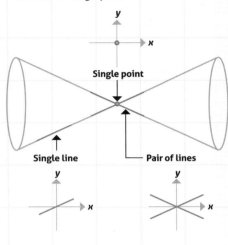

Single point

Single line · Pair of lines

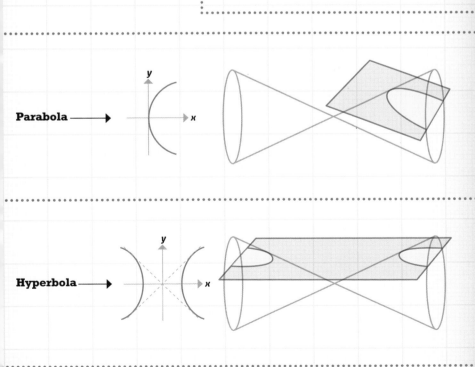

Parabola ⟶

Hyperbola ⟶

The Quadratic Formula

We've seen how a quadratic can be solved by "completing the square." Now we'll have a go at using the quadratic formula. The ruse is this: a couple of guys have dug a parabolic swimming pool that exends into both their backyards. The equation of the swimming pool is $6x2 + 5x - 21 = d$, where d is the depth below the ground and x is the number of meters away from the property line. How far does the pool extend into each property?

Areas of Concern

From the ancient Egyptians onward it has been necessary to work with problems involving areas, and quadratic equations have been known since this time.

However, it was with Eastern mathematics that it began to take on a more modern form. As we've already seen, Al-Khwarizmi had a number of methods for solving various quadratics, then Sridhara was one of the first to produce a general formula for solving quadratics. However, although his form of the quadratic formula was brought to Europe, it wasn't quite like the one we know today. We have to wait a few centuries until European mathematicians led by Girolamo Cardano (see pp. 92–93) began to work with a full range of solutions that included complex and imaginary numbers (see pp. 94–97).

In 1637, when René Descartes published *La Géométrie*, the quadratic formula had adopted the form we know today. It's worth noting that, while this book looks at two ways of solving quadratics, there are others you might want to explore.

The quadratic equation, or formula, is a generalized solution to any quadratic equation. It's really quite easy to use and the biggest mistakes tend to be simple errors in calculations rather than problems with understanding:

$$x = \frac{-b \pm \sqrt{b2 - 4ac}}{2a}$$

Where a, b, and c represent the coefficients in $ax^2 + bx + c = 0$.

So, for our problem: $a = 6$, $b = 5$, and $c = -21$ (don't forget the negative) after we plug in the variables, the formula becomes:

Here you can see a graph of the parabola that describes the swimming pool. You'll notice that we've made the values of the *d*-axis positive because we worked with positive values for depth rather than "negative height." This is simply to make the graph easier to understand.

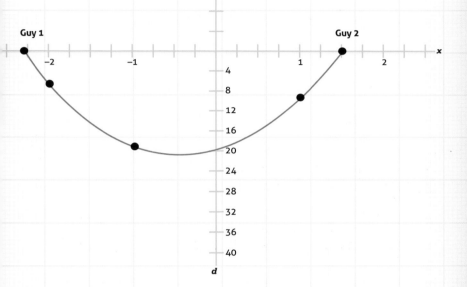

$$(6x^2 + 5x - 21 = d)$$

$$x = \frac{-(5) \pm \sqrt{(5)^2 - 4(6)(-21)}}{2(6)}$$

$$x = \frac{-5 \pm \sqrt{25 + 504}}{12}$$

$$x = \frac{-5 \pm \sqrt{529}}{12}$$

$$x = \frac{-5 \pm 23}{12}$$

Next we need to calculate both options:

$$x = \frac{-5 + 23}{12} \qquad x = \frac{-5 - 23}{12}$$

$$x = \frac{18}{12} \qquad x = \frac{-28}{12}$$

$$x = \frac{3}{2} \qquad x = \frac{-7}{3}$$

The solution is $x = \frac{3}{2}$ and $x = \frac{-7}{2}$ so the pool extends $\frac{7}{3}$ of a meter (2.33 m) into one guy's yard (I am assuming his house was to the left or negative side of the number line) and $\frac{3}{2}$ of a meter (1.5 m) into the other yard.

4

The
Italian
Connection

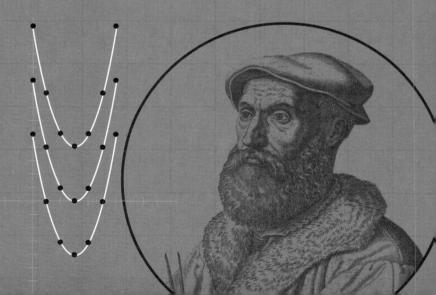

As the center of mathematics moved toward the West, Italian mathematicians played a crucial role in its development—both in bringing it to Europe in the first place, then developing it during the Renaissance. In this chapter we will meet such beautiful and creative mathematics as the Fibonacci sequence, revisit the golden ratio, and find ourselves in the strange company of imaginary and complex numbers.

Fibonacci Part 1

Although Archimedes, Gauss, and Newton are generally considered the "big three" of mathematics, there are two other mathematicians that are, in my opinion at least, more fun and accessible. These two mathematicians are Pascal, who we will meet in the next chapter, and Fibonacci, whose famous sequence we can see all around us.

African Roots

Fibonacci, was born in Pisa, Italy, in 1170. Although he was born in Italy, Fibonacci was actually raised and educated in North Africa. His father Guilielmo was a diplomat for the Republic of Pisa, representing merchants who traded through a port in what is now Algeria.

His education in what was then part of the Islamic empire introduced Fibonacci to a number system vastly superior to that which was used in Europe at the time. During his life, Fibonacci witnessed the fall of the Abbasid Dynasty. An-Nasir was the thirty-fourth Abbasid Caliph who reigned from 1180 to 1225, and is considered the last strong Abbasid Caliph. During this time most of Spain and Portugal were conquered by the Christians.

Fibonacci traveled widely until 1200, when he returned to Pisa. During his time there, Fibonacci wrote several texts including *Liber Abaci* (1202), *Practica Geometriae* (1220), *Flos* (1225), and *Liber Quadratorum* (1225). Fibonacci died in 1250 in Pisa.

Practica Geometriae and *Flos*

Practica Geometriae is eight chapters of geometry problems based on Euclid's *Elements* and *On Division of Figures*, one of which shows how to find the height of tall objects using similar triangles. In *Flos*, Fibonacci solves a cubic equation previously solved by Omar Khayyam (see pp. 72–73); though the solution is irrational, Fibonacci managed to get the answer correct to nine decimal places.

Liber Quadratorum

Liber Quadratorum (Book of Squares) is considered by some to be Fibonacci's best work, though it's not as famous as *Liber Abaci*. It's a text on number theory, so the practical applications aren't obvious, but the mathematics is fascinating nonetheless.

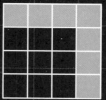

The first four square numbers. By observation we see that Fibonacci's idea that square numbers are the sums of odd numbers holds true.

1

4
(1 + 3)

9
(4 + 5)

16
(9 + 7)

In *Liber Quadratorum* Fibonacci looks at square numbers (see p. 16), among other things, and wrote that they are the sums of odd numbers. That is to say:

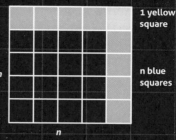

n blue squares

1 yellow square

n blue squares

$1 = 1$ (one is a square number)

$1 + 3 = 4$ (four is a square number)

$1 + 3 + 5 = 9$ (nine is a square number), and so on.

The diagram above right shows that this holds true for the first few square numbers, but it would be good if we could show it was true of any square number. If we have a large square that measures n by n and we add to the sides and top corner more squares measuring one-by-one we get a square measure $(n + 1)$ by $(n + 1)$. The number of one-by-one squares added is $2n + 1$, which is an odd number.

Therefore to increase for any square number n^2 to the next square number $(n + 1)^2$ you must add $n + n + 1$ or $2n + 1$. Therefore:

$$n^2 + 2n + 1 = (n + 1)^2$$

The important part is the addition of $2n + 1$ to the n^2. Given that n can be any natural number, $2n$ is guaranteed to be an even number because it's a multiple of two. This means that $2n + 1$ must be odd. If we start with $n = 1$, the first square number, we see from the formula that we will get successive odd numbers as n increases.

Fibonacci also shows how to find Pythagorean triples, which we met earlier (see p. 37). The first step is to take any odd square number as one of the shorter sides of a right-angled triangle. The other short side will be the sum of all the odd natural numbers up to the odd number chosen in the first step. Then you can add these two numbers together to complete a Pythagorean triple.

For example, take 25 as the first side. The sum of the odd natural numbers less than 25 is 144. Add 25 and 144 and you get 169, this can also be expressed as squares.

Let's Get Graphical

Solving quadratics (see pp. 60–61 and 76–77) may make some people break out in a cool sweat, but when plotted on a graph the form they take is really quite elegant. The parabola, as this graceful line is called, can be seen in many places in the real world; for example, the path a projectile follows is a parabola, if we ignore air resistance, and a satellite dish is the parabola that is perhaps closest to home.

Graphing the Parabola

The basic quadratic equation is $y = x^2$. The graph of this equation can be found by substituting a value for x so we can find the value of y. For example, when $x = -2$ the corresponding y value is 4 (in other words, $(-2)^2$, or $-2 \cdot -2$). This produces graph 1 (below).

From the bottom (vertex) of the parabola we go one unit left and right then up one to get two more points on the graph. Again from the vertex, we go out left and right two

and up four to get another two points. This squaring continues indefinitely. In all of these graphs we represent this by marking the changing values in the tables with points on the graph, but continuing the line to show that its shape remains the same.

Moving a parabola around a piece of graph paper is relatively easy. Let's look at the graphs of the basic parabola, $y = x^2$, and two others: $y = x^2 + 3$ and $y = x^2 - 3$. If we made a table of values for these three functions we would notice that the +3 and the –3 just increase and decrease

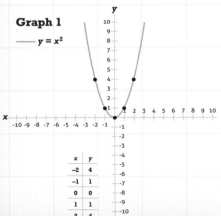

Graph 1

—— $y = x^2$

x	y
–2	4
–1	1
0	0
1	1
2	4

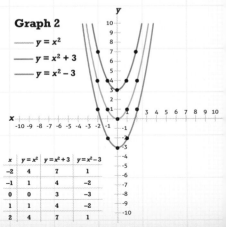

Graph 2

—— $y = x^2$
—— $y = x^2 + 3$
—— $y = x^2 - 3$

x	$y = x^2$	$y = x^2+3$	$y = x^2-3$
–2	4	7	1
–1	1	4	–2
0	0	3	–3
1	1	4	–2
2	4	7	1

Graph 3

— $y = x^2$

— $y = (x + 1)^2$

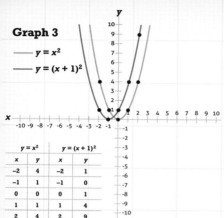

$y = x^2$		$y = (x + 1)^2$	
x	y	x	y
-2	4	-2	1
-1	1	-1	0
0	0	0	1
1	1	1	4
2	4	2	9

Graph 4

— $y = x^2$

— $y = (x - 1)^2$

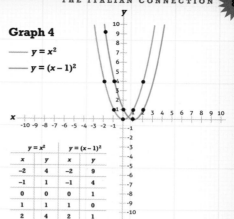

$y = x^2$		$y = (x - 1)^2$	
x	y	x	y
-2	4	-2	9
-1	1	-1	4
0	0	0	1
1	1	1	0
2	4	2	1

the y values and therefore move the parabola either up or down. That is to say that, given a graph of $y = x^2 \pm q$, the q value simply moves the parabola up or down (graph 2).

Moving the graph left and right is a little more complicated. Starting with the basic parabola, $y = x^2$, we will shift the curve left and right by adding and subtracting numbers within the square; for example, graph 3 (above left) shows the graph of our basic parabola $y = x^2$ and also the graph of $y = (x + 1)^2$, which is shifted to the left. The opposite occurs if we subtract values inside the bracket, for example, $y = (x - 1)^2$ in graph 4 (above right) shifts the graph to the right.

So if we have an equation in the form $y = (x \mp p)^2$, we know that the p moves the parabola left and right.

At this point a question usually arises: Why does the q variable do what it says, when the p variable does the opposite? That is to say, why does a positive q variable shift the curve in a positive direction (up); when a positive p variable shifts the curve in a negative direction (left)—and vice versa?

The answer is surprisingly straightforward. It's just to do with the way we like to write our equations. Really an equation of the form $y = x^2 - q$ should be written as $y + q = x^2$, but we tend to write our equations in the form "$y = ...$"

That said, the reason isn't too vital; it's just a question of getting your head around this quirk. Once you understand that, it's pretty simple to start drawing graphs of equations that are presented in the form $y = (x \mp p)^2 \pm q$. It's made even easier because the p and q variables don't affect each other—q just cares about up or down and nothing else, p cares about left and right and that's it. Just remember: positive q shifts up; positive p shifts left.

Fibonacci Part 2

Liber Abaci, not to be confused with the flamboyant pianist Liberace, is Fibonacci's most famous work. Written in 1202, it's to this work that Western math owes its renaissance. Through *Liber Abaci*, Fibonacci introduced Hindu-Arabic numerals to Europe, thus making arithmetic much easier.

The Modern Number System

The Hindu-Arabic number system has a long and storied past. Its beauty lies in the fact that it's a base-ten positional system (such as we use today) that allows for easy arithmetic. It began in India, and in the seventh century Brahmagupta formulated the first mathematical concepts that treat zero as a number, not just a place holder.

These ideas flowed westward through the Islamic empire. They were passed on by Al-Khwarizmi in the early ninth century and then on to Fibonacci. Though *Liber Abaci* was not the first text to introduce Hindu-Arabic numerals to Europe, it was the first to catch on. This can be attributed to the practical nature of its presentation and the benefits Fibonacci saw in the system. The first section dealt with the arithmetic of the Hindu-Arabic system and the second section with problems confronted by merchants.

> Though *Liber Abaci* was not the first text to introduce Hindu-Arabic numerals to Europe, it was the first to catch on.

Although *Liber Abaci* is known for introducing Hindu-Arabic numerals and the Fibonacci sequence, it was also directed at merchants and the finance of trade.

The Fibonacci Sequence

....

In the third section of the book, a problem dealing with rabbits introduced what would become known as the Fibonacci sequence. The problem went something like this:

A man starts with a pair of bunnies (one male, one female). It takes one month for them to mature to the point when they can breed. Then it takes one month for impregnation, pregnancy, and birth. Each birth produces one male and one female. How many pairs of rabbits will you have at the end of each month?

To start, the man has one pair. At the start of the next month he will still have one pair, but they will be mature. At the start of month three he will have two pairs: the original and the newborn pair. Note that the original pair will begin to produce again while the newborn pair needs one month to grow up. At the start of the fourth month he will have three pairs: the original, the first generation (which is ready to breed), and a new pair of bunnies. At the start of month five we have five pairs of rabbits because there are now two pairs giving birth.

End of month

Number of pairs

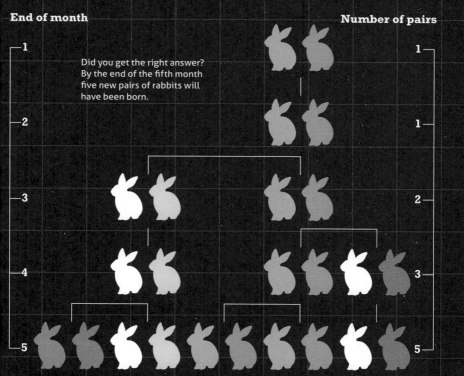

Did you get the right answer? By the end of the fifth month five new pairs of rabbits will have been born.

Mathematically the formula for finding the numbers in a Fibonacci sequence is $f_{n+2} = f_{n+1} + f^n$ which looks weird but just means that the new Fibonacci number (f_{n+2}), is the sum of the two preceding Fibonacci numbers $f_{n+1} + f_n$. The first few terms in the sequence are: 1, 1, 2, 3, 5, 8, 13, 21, 34, 55, 89, and so on.

Each number of the Fibonacci sequence is the sum of the two numbers that precede it.

The segmented head of a sunflower, rendered as dots.

1 1 2 3 5 8 13 21 34 55 89 144 233 377 610 987 1597 25

Numbers in Nature

Natural examples of the Fibonacci sequence are many, including in flowers (see below). One of the more spectacular instances can be found in the head of a sunflower, whose beautifully segmented center is arranged into Fibonacci spirals. Other naturally occurring objects, such as shells and coral, also take this form.

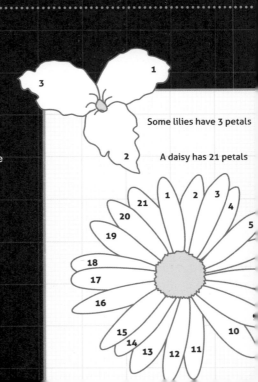

Some lilies have 3 petals

A daisy has 21 petals

6765 10946 17711 28657 46368 75025 121393 196418 317811

The number of petals on a
sunflower should be 34, 55, or 89,
depending upon its size.

Natural examples of Fibonacci
spirals include this beautifully
patterned romanesco broccoli.

THE FIBONACCI SEQUENCE IN PETALS

Fibonacci numbers can be found all through nature. There are many flowers
that have petals which match Fibonacci numbers. Here is a short list:

A wild rose
has 5 petals

3 petals: lily, iris

5 petals: buttercup, wild rose, larkspur, columbine (aquilegia)

8 petals: delphiniums

13 petals: daisies, ragwort, corn marigold, cineraria

21 petals: daisies, aster, black-eyed susan, chicory

34 petals: daisies, plantain, pyrethrum

55, 89 petals: michaelmas daisies, the *Asteraceae* family

The Golden Ratio

Some numbers are just cool. Earlier I wrote about my π shirt, but there are plenty of other neat numbers as well. Here we're going to take a look at phi (φ), which is one of the most beautiful and fascinating numbers of all.

The Golden Ratio

The value of phi (φ), also known as the golden ratio or section, is $\frac{1+\sqrt{5}}{2}$. This might seem like a weird number, but like the other cool numbers, it tends to pop up all over the place. Some say that its appearance is just a coincidence and that if you look for something you're sure to find it. I think these people miss out on the joys of life.

Like π, φ is an irrational number, meaning that if you try to write it as a decimal you will be busy for a very long time indeed! However, just to give the first handful of

The rectangle below (ACFD) is a golden rectangle; the ratio of its sides AC/CF is the golden ratio. If we remove the square ABED, a new rectangle is formed (BCFE), which is also golden. If we remove the square BCHG, we find yet another golden rectangle (GHFE), and so on. Drawing a curve from corner to opposite corner of each box creates a "golden spiral."

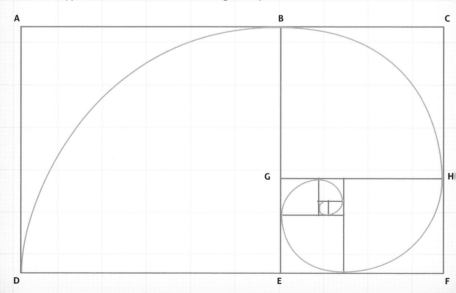

decimal places, the value of the golden ratio is roughly 1.618033989.

It's here that things get interesting, as when we chart the Fibonacci numbers and divide one number by its immediate predecessor, the result starts to approach the golden ratio.

f_n	$f_n \div f_{n-1}$
1	N/A
1	1
2	2
3	1.5
5	1.666667
8	1.6
13	1.625
...	...

Finding the Golden Ratio

Using the Fibonacci sequence is just one way to find the value of φ. The golden ratio is $\frac{1+\sqrt{5}}{2}$, but it can also be written as a repeating fraction (the three dots mean that the fractions go on forever).

$$1 + \cfrac{1}{1 + \cfrac{1}{1 + \cfrac{1}{1 + \cfrac{1}{\ddots}}}}$$

We can see the convergence to φ by ignoring parts of the repeating fraction and just using the beginning parts.

A nautilus shell is a great example of the golden ratio in nature.

First term = 1

Second term = 1+1 = 2

Third term = $1 + \dfrac{1}{1+1} = 1 + \dfrac{1}{2} = 1.5$

Now this could get torturous (some might say it already has), but it's easier if we realize the denominator is just the previous term. So the fourth term is:

$$1 + \frac{1}{\text{third term}} = 1 + \frac{1}{1.5} = 1.6$$

If we continued we'd find ourselves again converging on the golden ratio. And the same is true when φ is expressed as a repeating square root.

$$\varphi = \sqrt{1 + \sqrt{1 + \sqrt{1 + \sqrt{1 + \ldots}}}}$$

Yet another interesting property of φ is that its reciprocal (one divided by φ, or $\frac{1}{\varphi}$, is equal to one less than φ, or $\frac{1}{\varphi} = \varphi - 1$).

The Golden Ratio in Life

The golden ratio can be found throughout nature. In the human body, given the slightly unreasonable assumption that you have the perfect figure, we find the ratio in: your overall height divided by the height to your navel; the ratios of the lengths of the bones in your fingers; the ratio of the length from your elbow to your wrist to the length of your hand.

In fact, φ, or the golden ratio, shows up wherever the Fibonacci sequence can be seen. There are many examples of it in nature, and it can also be found in the art and architecture of humankind; for example, the proportions of the *Mona Lisa* or the Parthenon in Athens.

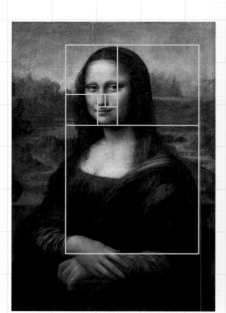

Famously, the compostion of Leonardo da Vinci's *Mona Lisa* is said to follow the golden ratio.

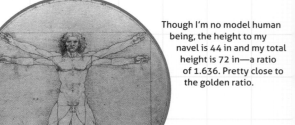

Though I'm no model human being, the height to my navel is 44 in and my total height is 72 in—a ratio of 1.636. Pretty close to the golden ratio.

In one final example, the golden ratio also can be found in the lines of a pentagram—the ratios of certain lengths, for example, $\frac{AD}{AC}$, $\frac{AC}{AB}$, and $\frac{AB}{BC}$ are all the golden ratio. The same relationship can be found in other shapes, as shown below.

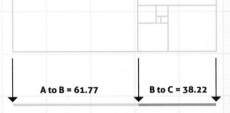

A to B = 61.77   B to C = 38.22

The ratio between AB and BC is the golden ratio.

The pentagram is filled with golden ratios.

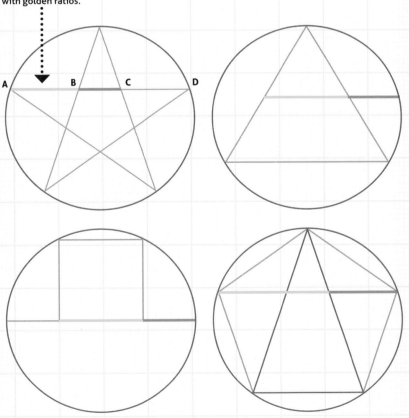

Tartaglia & Cardano

Fast-forward some two hundred and fifty years and quadratics have become yesterday's news. Mathematicians had more or less got their heads around them and mathematics had moved on. In Italy, in the 1500s, cubic equations were where it was at. Today cubic equations are involved in all sorts of applications, particularly those that involve volumes.

Niccolo Tartaglia

Niccolo Fontana Tartaglia was born in around 1500 in Brescia, Italy, which at the time was part of the Republic of Venice. Due to Tartaglia's ability in mathematics, his mother found him a patron and he went to study in Padua, Italy, eventually teaching mathematics in Verona and Venice. From this time until his death in 1557, Tartaglia was involved in a series of arguments with other Italian mathematicians.

Girolamo Cardano

Girolamo Cardano was born in Pavia, Northern Italy in 1501. His father Fazio was a lawyer but because of his skills in mathematics he lectured on geometry at the University of Pavia and the Piatti Foundation in Milan. Cardano's father also consulted with Leonardo da Vinci on problems in geometry.

Cardano's first job was that of an assistant to his father. He entered medical school in Pavia and was very successful, although he made many enemies due to his abrasive demeanor. He received his doctorate of medicine in 1525, but his outspoken nature made it difficult for him to be admitted to the College of Physicians in Milan and it would be fourteen years before he gained entry.

He married in 1531, and during the 14 years between his degree and his admission to the College of Physicians in Milan, Cardano worked as a physician in a small village and lecturing mathematics at the Piatti Foundation. In 1539 Cardano started his correspondence with Tartaglia, and the rest of his life was filled with controversy.

It wasn't just mathematical problems that engulfed Cardano's life. His oldest son was convicted of the murder of his wife and executed. His youngest son was a gambler, and not a good one—he gambled away most of his possessions and much of Cardano's money. Then in 1570 Cardano was accused of heresy for having published a horoscope of Jesus over a decade earlier and was put in jail for several months. Cardano moved to Rome, where he died in 1576.

Tartaglia Takes on the World

· · · · · · · · · · · · · · · ·

Scipione del Ferro, an Italian mathematician who was born in Bologna in 1465, is credited with being the first to solve cubic equations. He kept his process or formula for solving these equations under wraps until 1526 when, on his deathbed he revealed his secret to his student, Antonio Fior.

Fior boasted that he could solve cubics and after some mathematical "trash talk" he and Tartaglia set about what amounted to a mathematical duel. Each would give the other thirty problems to be solved over a period of time. To kick things off, Fior gave Tartaglia thirty problems of the form $x^3 + ax = b$. Tartaglia, however, knew how to solve equations of the form $x^3 + ax^2 = b$. After much effort, Tartaglia

eventually solved them all. He was now able to solve cubics in both forms. The problems Tartaglia set in return were more varied and Fior lost.

At this point Cardano became interested in cubic equations and wrote repeatedly to Tartaglia to ask for his method. Tartaglia was eventually persuaded to share the information, on the condition that Cardano never reveal it to anyone else and would only keep the information in code.

However, in 1543 Cardano discovered that del Ferro was actually the first to solve cubic equations, and no longer felt obliged to honor his promise to Tartaglia. Consequently, he published *Ars Magna* in 1545, which included methods for solving cubic equations both from del Ferro and Tartaglia, as well as the advances that Cardano and Ferrari had made themselves. This established Cardano's reputation as a leading mathematician.

In 1570 Cardano was accused of heresy for having published a horoscope of Jesus over a decade earlier, and was put in jail for several months.

Imaginary Math

New number types have caused controversy throughout the ages. As new problems are discovered, their solutions often require new mathematics, and with this new mathematics come new number types. Imaginary and complex numbers are great examples of this; but before we look at them more closely let's review the development of our number system.

Imaginary numbers, and by extension complex numbers, have many applications. Complex numbers are needed when working with electromagnetic fields and alternating current (AC) circuits. Quantum mechanics and even those cool fractal posters require complex numbers as well. Control systems also require complex numbers.

Controversy in the Past

During the time of Pythagoras we had the natural numbers and we had positive fractions or rational numbers. Then, when the Pythagoreans discovered irrational numbers people got quite upset. The idea of a number that can't be expressed as a fraction was just too dangerous.

Now, however, we see irrational numbers as an important part of mathematics—how else would you solve $x^2 = 2$? And eventually the Greeks became accustomed to the idea.

Zero also caused problems over the ages. It should be pointed out that we're talking about zero the number not zero the place holder—yes, there is a difference. For many years the idea of zero as a place holder existed, but not as a number itself. To the Greeks, who looked upon mathematics from a geometric perspective, zero seemed absurd. When numbers or unknowns represented lengths, and squares represented areas, zero had no place. Why solve a problem that did not exist? If a length is zero there is no

The Mandelbrot set is a fractal generated by a complex quadratic, but all that really matters is how great it looks!

line; if an area is zero there is no object. It was Brahmagupta (see pp. 64–65) who tried to place zero within the rules of arithmetic.

Acceptance of negative numbers took even longer—zero at least had the advantage of starting out as a place holder. Again it was Brahmagupta who tried to bring negative numbers into the family. European mathematicians continued to have trouble with them even during the sixteenth century, when Italian mathematicians were starting to contemplate imaginary numbers.

Imaginary and Complex Numbers
· ·

Let's begin by saying that "imaginary" is a bad label for this type of number. It implies that these numbers do not exist when they do. In fact, they are very real and very necessary in mathematics.

An imaginary number is just $\sqrt{-1}$ and is represented by the letter i or j—mathematicians use i while engineers tend to use j. The use of imaginary numbers helps us solve some simple-looking equations. If we had $x^2 - 1 = 0$ we could solve this by isolating x^2 and get $x^2 = 1$ and then $x = \pm 1$. This type of equation is easy to solve. If we change the problem to $x^2 + 1 = 0$ and we isolate x^2, then we get $x^2 = -1$. But what number, when squared, equals a negative number? Well, nothing, unless you discover a new number. This is how the imaginary number was born. If we make $i = \sqrt{-1}$, then

$i^2 = -1$. It follows then that the solution to the problem $x^2 + 1 = 0$ is $x = \pm i$. It doesn't matter if you don't understand right now, complex arithmetic is covered on pages 96–7.

Complex numbers, meanwhile, are numbers with a real and imaginary component. For example, $3 + 4i$ is a complex number because the 3 can be considered a "real three" while the $4i$ is an imaginary number.

A LITTLE HISTORY

Just like irrational numbers, zero, and negative numbers, imaginary and complex numbers have caused their own controversy over the years. The first published account of complex numbers comes from Cardano's *Ars Magna*. When solving cubic and quartic equations, Cardano came across the square root of a negative number in the middle of the calculation. Ignoring the fact that this was an "imaginary" or "impossible" situation he continued the calculation to produce a "real" result.

Rafael Bombelli (see p. 101) was the first to work explicitly with complex numbers and wrote of operations with them in 1572. René Descartes (see pp. 106–107) is sometimes credited with giving imaginary numbers their name, and later Carl Gauss (see pp. 128–29) introduced the term "complex number."

Complex Arithmetic

Complex arithmetic is not that complex really. In fact, all it requires is an understanding of the Cartesian coordinate system—graph paper to you and me—and a little trigonometry. It also requires that you accept the existence of imaginary numbers; if you can separate numbers from the physical world then that helps. There are, however, masses of real-world applications for complex numbers, including alternating-current (AC) circuits.

The Real Number Line

Let's start with the real number line, our old friend from when we were learning to count and do basic addition and subtraction. Assuming we placed a frog at 4 on the number line then multiplied by −1, the frog would hop all the way across the number line and land on −4. As the frog turned all the way around we can see that the angle of the hop was 180°. If we then multiply by −1 again we end up back at 4; that's another 180° turn, making a total of 360°. In both cases multiplying by −1 causes a hop of 180°.

Now we're going to add another axis to our number line. The horizontal axis represents the real numbers and the vertical axis represents the imaginary numbers (see above right).

Since $i = \sqrt{-1}$ we can think of i as half a negative sign. If we start our frog at 4 again and multiply by i, this moves the frog to $4i$ on the graph— the frog has hopped 90°, half the value it did for a multiplication by −1. If we then take this position $4i$ and multiply by i again, we get $4i^2$.

Multiplying by −1 causes the frog to hop through 180°.

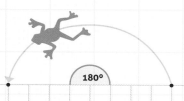

If $i = \sqrt{-1}$ then $i^2 = -1$ and $4i^2$ is equal to -4. This represents another hop of 90°. Therefore, when we multiply by -1 we hop 180° and when we multiply by i we hop 90°. If our frog started at 4 on the number line, and is then multiplied by i three times (or i^3) it would hop through 270° and end up at $-4i$. This is because two of the three is made the negative sign and left one extra i.

Complex Numbers

Complex numbers are numbers that have both real and imaginary components. To represent these numbers we can place them on the complex plain (our graph paper with a real and imaginary axis). If we have the number $(3 + 4i)$, it would be a point three to the right and four up. If we draw a line from the origin (where the vertical and horizontal lines meet) to the point, we can determine the length of the line using the Pythagorean theorem (see pp. 36–37) and the angle the line makes with the positive real axis using trigonometry (see pp. 28–31). Using the Pythagorean theorem $a^2 + b^2 = c^2$, where a and b are 3 and 4, we get $3^2 + 4^2 = c^2$ then $9 + 16 = c^2$ then $25 = c^2$ or $c = 5$. This represents the absolute value or modulus of the complex number. To find the angle we take the inverse tangent of $\frac{4}{3}$ or $\tan^{-1}\left(\frac{4}{3}\right) = \theta \approx 53°$.

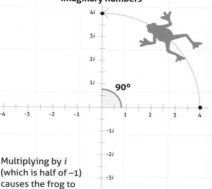

Imaginary numbers

Multiplying by i (which is half of -1) causes the frog to hop through 90°.

Multiplying a Complex Number by i

Earlier we said that multiplying by i causes a rotation (or hop) of 90°. To demonstrate this let's multiply $(3 + 4i)$ by i. Given $i(3 + 4i)$, the i swoops in to give $3i + 4i^2$. As $4i^2$ is -4, the new complex number is $-4 + 3i$. This number is now in the upper left part of the graph paper.

Again, using the Pythagorean theorem, we find that the length of the line is five and using trigonometry we find the angle the line makes with the negative real axis. To find the angle we can take the inverse tangent of $\frac{3}{4}$ or $\tan^{-1}\left(\frac{3}{4}\right) = \theta \approx 37°$. If we look at the angle formed between the number $(3 + 4i)$ and $(-4 + 3i)$ we see that it's 90°. So multiplying by i rotates the point (line) by 90° but does not change the length. The next section will deal with multiplying and adding complex numbers.

Adding Complex Numbers
....

Adding complex numbers is quite easy; you just add the real parts together and then the imaginary parts together. For example, let's add $(3 + 4i) + (2 + 5i)$. This becomes $(5 + 9i)$.

Graphically we can look at this in two ways. In the first we draw the line for each complex number from the origin. The first number goes to the right three and the second number goes to the right two—the total distance is five. The first number goes up four and the second goes up five—the total distance is nine. The second approach is to use a tip-to-tail method (see graph 1). From the tip of the first number you place the tail of the second. So first you move to three right and four up, then from this point you move another two right and five up to finish five right and nine up.

Graph 1

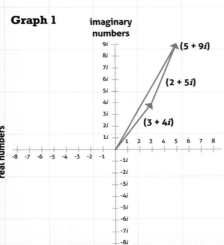

Multiplying Complex Numbers
....

Multiplying complex numbers is no different from multiplying binomials (see pp. 122–23); complex numbers are just binomials with a real and imaginary part. As an example, let's multiply $(3 + 4i)(2+5i)$:

$(3 + 4i)(2+5i) = 6 + 15i + 8i +20i^2 = 6 + 23i - 20 = -14+23i$

Remember that $i^2 = -1$ so $20i^2$ is equal to $20(-1)$ or -20.

This can be demonstrated graphically (connecting the geometry to the algebra) by plotting the two binomials and the resulting binomial. We will do each step individually.

Graphing $(3 + 4i)$
....

If we plot $(3 + 4i)$ we have a point three to the right and four up. If we make a line from the origin and use the Pythagorean theorem, we find the length of this line is five:

$3^2 + 4^2 = c^2$ then

$9 + 16 = c^2$ then

$25 = c^2$ or $c = 5$.

The angle can be found by using trigonometry (see p. 29); we take the inverse tangent of $\frac{4}{3}$ or:

$\tan^{-1}\left(\frac{4}{3}\right) = \theta \approx 53.13°$

Graphing (2 + 5*i*)

If we plot (2 + 5*i*) then we have a point that is two to the right and five up. If we make a line from the origin and use the Pythagorean theorem we find that the length of this line is $\sqrt{29}$ or approximately:

$2^2 + 5^2 = c^2$ then

$4 + 25 = c^2$ then

$29 = c^2$ or

$c = \sqrt{29}$ or $c = 5.3852$

The angle can be found by using trigonometry; we take the inverse tangent of $\frac{5}{2}$ or $\tan^{-1}\left(\frac{5}{2}\right) = \theta \approx 68.20°$.

Graphing (–14 + 23*i*)

If we plot (–14 + 23*i*) we have a point fourteen to the left and twenty-three up. If we make a line from the origin and use the Pythagorean theorem we find the length of this line is $\sqrt{725}$ or approximately 26.926:

$(-14)^2 + 23^2 = c^2$ then

$196 + 529 = c^2$ then

$725 = c^2$ or

$c = \sqrt{725}$ or $c = 26.926$

The angle can be found by using trigonometry; we take the inverse tangent of $\frac{23}{14}$ or $\tan^{-1}\left(\frac{23}{14}\right) = \theta \approx 58.67°$.

This angle is the angle between the negative real axis and the line.

To find the angle between the line and the positive real axis we subtract this angle from 180°. The angle from the positive real axis is 180° – 58.67° or 121.33°.

Bringing it Together

This length of the resultant binomial 26.926 represents the multiplication of the lengths of the two original binomials 5 and 5.3852. The angle of the resultant binomial (121.33°) is the sum of the angles that the two original binomials formed with the positive real axis. To view this graphically (see Graph 2, below), when you multiply two complex numbers you multiply their lengths and you add the angles they form with the positive real axis.

Complex number	Length	Angle
(3 + 4*i*)	5	53.13°
(2 + 5*i*)	5.3852	68.20°
(–14 + 23*i*)	26.926	121.33°

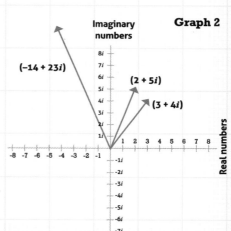

Graph 2

Conjugates

....

Conjugates are a neat way to produce a real number from two complex numbers. Conjugates are complex numbers where the real parts are the same but the imaginary parts are negatives of each other. For example, $(3 + 4i)$ and $(3 – 4i)$ are conjugates. They are helpful because when you multiply conjugates the imaginary part disappears.

Doubly Complex

....

Let's start with the following expression:

$$(3 + 4i)(3 – 4i)$$

Which we can "foil" to give:

$$9 – 12i + 12i – 16i^2$$

Remember that $i^2 = –1$ so $–16i^2$ is equal to $–16(–1)$ or $+16$, and we can cancel out the $–12i$ and the $12i$ to give:

$$9 + 16 = 25.$$

When we multiply complex numbers we multiply the lengths and we add the angles (see p. 98). Looking at Graph 1 we can see that the first complex number has an angle of 53.13° above the positive real axis and the second complex number has an angle of 53.13° below the positive real axis. Adding these angles therefore gives 0° and produces a real number.

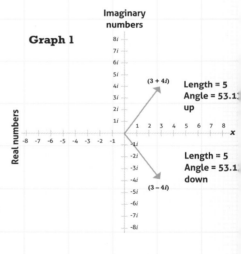

Graph 1

Length = 5
Angle = 53.13
up

Length = 5
Angle = 53.13
down

When solving polynomial equations with real coefficients, the solution sometimes includes complex numbers. When this happens the complex numbers come in pairs: one complex number and its conjugate.

Conjugates Put to Work

To divide complex numbers is a little more complicated and requires the use of the conjugate. As an example, let's divide $\frac{(-8 + 3i)}{(3 + 2i)}$. First we multiply the top and bottom by the conjugate of the bottom, then "foil" out the top and bottom, then simplify, this gives a chain of equations as follows:

$$\frac{(-8 + 3i)}{(3 + 2i)} = \frac{(-8 + 3i)}{(3 + 2i)} \cdot \frac{(3 - 2i)}{(3 - 2i)} = \frac{-24 + 16i + 9i - 6i^2}{9 - 6i + 6i - 4i^2} =$$

$$\frac{-24 + 25i + 6}{9 + 4} = \frac{-18 + 25i}{13} = \frac{-18}{13} + \frac{25i}{13}$$

Division follows the same graphical approach as multiplication. But when dividing complex numbers you divide their lengths and subtract their angles.

so $\frac{(-8 + 3i)}{(3 + 2i)} = \frac{-18}{13} + \frac{25i}{13}$

RAFAEL BOMBELLI

Rafael Bombelli was born in Bologna, Italy, in 1526. Following on from Cardano and Tartaglia, he and Lodovico Ferrari, Cardano's assistant, represented the next generation of great mathematicians from northern Italy—the hub of mathematics at the time. Bombelli didn't receive a university education; instead he learned his mathematics from Pier Francesco Clementi, an architect and engineer.

Bombelli followed Clementi into the field of engineering and began working on land-reclamation projects. But in 1555, when the project he was working on was suspended, Bombelli decided to write a comprehensive review of algebra with the aim of making the subject more accessible. However, Bombelli's day job recommenced in 1560, before he could finish his book. It would be nearly a decade before his writings would be published. However, this wasn't necessarily such a bad thing.

Bombelli was invited to Rome to work on further engineering projects, and during his time there he was introduced to the work of the Greek mathematician Diophantus (see pp. 54-55). Bombelli undertook to make a translation of Diophantus's *Arithmetica*, but although the work remained unfinished it greatly influenced his work on algebra. When it was eventually published in three parts, Bombelli's *Algebra* included a number of problems he had taken from Diophantus. He intended to publish two further parts on geometry, but these remained unfinished when he died in 1572, although their manuscripts have since been discovered.

Bombelli's work was significant for two reasons: first for his comfortable nature working with negative numbers; and second for laying down the rules of addition, subtraction, and multiplication of complex numbers.

Quadratics, Parabolas, and Complex Numbers

Quadratics are very important. Besides their use in expressing the force that keeps us all aboard our giant space marble (gravity), quadratics are useful in control systems everywhere, from pulp mills to chemical plants.

It All Comes Together

In chapter three we met quadratics and looked at one of the ways they can be solved. Then, earlier in this chapter, we graphed parabolas and met complex arithmetic for the first time. Now we bring these ideas together.

When solving polynomials, in this case quadratics, we're looking for the value of x which makes the equation equal to zero. When we graph the quadratics (parabolas) we introduce the y variable to construct the graph.

A Quadratic with Two Solutions

To start, let's use a quadratic which would work nicely for mathematicians through the ages—positive whole number solutions. Let's analyze $0 = x^2 - 6x + 5$ in two different ways: by looking at the graph and by using the quadratic formula.

If we use the quadratic formula then we get:

$$\frac{-b \pm \sqrt{b^2 - 4ac}}{2a}$$

$$\frac{-(-6) \pm \sqrt{(-6)^2 - 4(1)(5)}}{2(1)}$$

$$\frac{6 \pm \sqrt{36 - 20}}{2}$$

$$\frac{6 \pm \sqrt{16}}{2}$$

$$\frac{6 \pm 4}{2}$$

which becomes $\frac{(6+4)}{2} = \frac{10}{2} = 5$ and $\frac{(6-4)}{2} = \frac{2}{2} = 1$.

So the solution to the polynomial equation $0 = x^2 - 6x + 5$ can be found by using the quadratic formula. These solutions also represent where the graph of $y = x^2 - 6x + 5$ crosses the x-axis.

A Quadratic with One Solution

Now let's analyze $0 = x^2 - 6x + 9$. If we use the quadratic formula we get:

$$\frac{[-\,b\pm\sqrt{b^2-4ac}}{2a}$$

$$\frac{-(-6)\pm\sqrt{(-6)^2-4(1)(9)}}{2(1)}$$

$$\frac{6\pm\sqrt{36-36}}{2}$$

$$\frac{6\pm\sqrt{0}}{2}$$

$$\frac{6\pm0}{2}$$

which becomes $\frac{(6+0)}{2}=\frac{6}{2}=3$ and $\frac{(6-0)}{2}=\frac{6}{2}=3$.

So the solution to the polynomial equation $0=x^2-6x+9$ can be found using the quadratic formula. These solutions also represent where the graph of $y=x^2-6x+9$ meets the x-axis. In this case there are two equal solutions, so the graph just touches the x-axis.

A Quadratic with Imagination

Now let's analyze $0=x^2-6x+13$. If we use the quadratic formula we get:

$$\frac{-b\pm\sqrt{b^2-4ac}}{2a}$$

$$\frac{-(-6)\pm\sqrt{(-6)^2-4(1)(13)}}{2(1)}$$

$$\frac{6\pm\sqrt{36-52}}{2}$$

$$\frac{6\pm\sqrt{-16}}{2}$$

$$\frac{6\pm4i}{2}$$

which becomes $\frac{(6+4i)}{2}$ or $(3+2i)$ and $\frac{(6-4i)}{2}$ or $(3-2i)$.

So the solution to the polynomial equation can be found by using the quadratic formula. Because the graph does not cross the x-axis, there are no real solutions. The solutions now have an imaginary component and are complex. Notice that the solutions $(3+2i)$ and $(3-2i)$ are conjugates.

To Summarize

When the value under the square root (called the discriminant) is positive, the equation has two distinct real solutions, so the parabola crosses the x-axis in two places. When the value of the square root is zero, the equation has two equal solutions, so the parabola just touches the x-axis. When the square root is negative, the solutions are complex numbers so the parabola does not touch the x-axis at all.

Graph 1

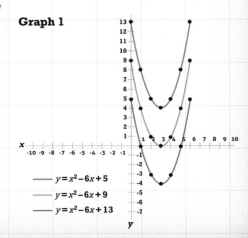

$y=x^2-6x+5$
$y=x^2-6x+9$
$y=x^2-6x+13$

Post-Renaissance Europe

As the Renaissance began to spread from Italy, Europe came alive with mathematical creativity. It was during this early modern period that some of the greatest mathematicians of all time lived: Pascal, Descartes, and Gauss. We shall meet some of these geniuses in this chapter, and with them one of the most beautiful mathematical constructions of all: Pascal's triangle.

René Descartes ◄

If you try to look for the town of La Haye in France, you will not find it. In its place you will find the town of Descartes. In 1802 La Haye changed its name to La Haye-Descartes in honor of René Descartes; then, in 1967 it dropped the La Haye altogether and simply became known as Descartes. To have a street named after you is a big deal, but to have your hometown change its name, that's something else.

A Student of Law

Descartes was born in La Haye in 1596. When he was a baby his mother died of tuberculosis. At the age of eight he entered a Jesuit college in La Flèche, where he studied until the age of sixteen. He received a degree in law from the University of Poitiers in 1616 and shortly after this, he joined the army.

One story has it that in 1619, while walking through the streets of Breda, in the Netherlands, he came upon a poster written in Dutch and asked a passerby, in Latin, to translate it for him. The passerby was Isaac Beeckman, a Dutch philosopher and scientist some eight years Descartes' senior. Beeckman agreed to translate the poster, which was a geometry problem, if Descartes would be willing to solve it. Of course, Descartes solved the problem in just a few hours and thus began a long friendship.

In the spring of 1621, around his twenty-fifth birthday, Descartes resigned from the army, and traveled throughout Europe. His travels took him to Bohemia, Hungary, Germany, Holland, France, then back to Holland in 1628.

"Of all things, good sense is the most fairly distributed: everyone thinks he is so well supplied with it that even those who are the hardest to satisfy in every other respect never desire more of it than they already have."

—René Descartes

MEDITATIONS ON FIRST PHILOSOPHY:
Expands on the work in *Discourse on Method* regarding the mind and body, truth and error, and existence.

PRINCIPLES OF PHILOSOPHY:
In this work Descartes tries to present the universe from a mathematical standpoint.

PASSIONS OF THE SOUL:
This work, which was dedicated to Princess Elizabeth of Bohemia, deals with emotions.

Descartes in Holland

It was in Holland that Descartes produced the works that made him famous among both mathematicians and philosophers. Shortly after arriving there, Descartes was writing a book, *Le Monde (The World)*, but decided not to publish it. Why? He had just heard of Galileo's subjection to house arrest in Italy for daring to challenge the Church's view of the universe. Descartes' next work was *Discours de la Méthode Pour Bien Conduire sa Raison et Chercher la Verité dans les Sciences* or *"Discourse on the Method of Reasoning Well and Seeking Truth in the Sciences,"* better known as *Discourse on Method*, which was published in 1637.

Discourse on Method

It's in this work, which examines how we can really know the world, that we can find what is probably the most famous quote in philosophy: *Cogito ergo sum*, or "I think, therefore I am," originally written as "Je pense donc je suis."

Discourse on Method has three appendices: "La Dioptrique," a work on optics, "Les Meteores," a work on meteorology, and "La Geometrie," a work on geometry.

"La Geometrie" is the most significant part of *Discourse on Method*. It's in this appendix that Descartes lays out the framework of analytic geometry. From this text we get the basis of our work in algebra, and it's at this point that students of today can read a work of algebra and not confront problems with notation. Descartes created a connection between geometry and algebra that we now take for granted. It's also interesting to note that it's from this appendix that we get the Cartesian coordinate system (graph paper to you and me). (Cartesian simply means relating to Descartes, so it wasn't just a town that he had named after him!)

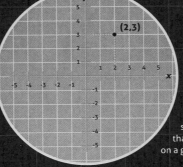

Descartes invented the system of coordinates that we use to plot points on a graph.

Graphing Lines

A graph gives us a picture of an equation, and a way to see what the numbers mean. The most common mathematical relationships are linear; for example, phone charges and minutes spent calling, or distance traveled at a constant speed over time. With so many applications, there is a great emphasis on linear functions (the lines of a graph), and there are many forms a linear function can take, with many different equations to represent a line. Here are three commonly used forms.

Slope-Point Form

For this first example we will give the information about the line, then find the equation and graph the line. The equation in slope-point form is: $y - y_1 = m(x - x_1)$, where x_1 and y_1 represent the point on the line and m represents the slope of the line.

To help, let's do an example. A line passes through the point (3, 4) and has a slope of $\frac{2}{3}$. Write the equation and graph the line.

Well, (3, 4) is the point and in the formula is represented by x_1 and y_1, while m represents the slope $\frac{2}{3}$. So the equation is $y - 4 = \frac{2}{3}(x - 3)$.

To graph the line we go to the point on the graph paper and plot it. Next, $\frac{2}{3}$ is the slope which represents the rise and run of the line ($\frac{2}{3} = \frac{\text{rise}}{\text{run}}$). From the point (3, 4) we rise two units and run right

three units. Lastly, we draw a line through the points to complete the graph below.

Slope y-Intercept Form

In this second example we have a graph of the line, and have to find the equation and information for the line. The equation in slope y-intercept form looks like $y = mx + b$, where m

Graph 1

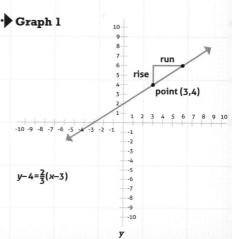

$y - 4 = \frac{2}{3}(x - 3)$

is the slope and b is the y-intercept (in other words, where the graph crosses the y-axis). Let's find the slope and y-intercept of graph 2 then write the equation.

From the graph we see that the line crosses the y-axis at the point 3. This represents the y-intercept, or b. From the y-intercept we see that the line drops one unit then moves right two units. This represents a slope of $\frac{-1}{2}$, therefore $m = \frac{-1}{2}$. Given this information the equation will look like: $y = \frac{-1}{2}x + 3$.

Standard Form

In this example, we have the equation and from it can graph the line and determine the information about it. An equation in standard form looks like $Ax + By = C$ where A, B, and C are integers and A must be positive. (The restrictions on A, B, and C are arbitrary.) As an example, let's graph $2x - 3y = 12$.

To graph this line let's use what I call the "cover up" method. The x and y axes are often overlooked, but when you have a point on the y-axis, you know one thing for sure, the x value of that point. The x value of any point on the y-axis is zero. This is very helpful. If we make $x = 0$ in the equation we will find a point on the y-axis. So we "cover up" the x term and solve the equation $-3y = 12$. By isolating y we get $y = -4$ which is the y-intercept.

We use the same approach to find the x-intercept: "cover up" the y term and get the equation $2x = 12$ or $x = 6$, which is the x-intercept.

We now have two points, through which we can draw a line. Now that we have the line we can state information about it. We already know the x and y intercepts and from them we can find the slope of $\frac{4}{6}$ or $\frac{2}{3}$.

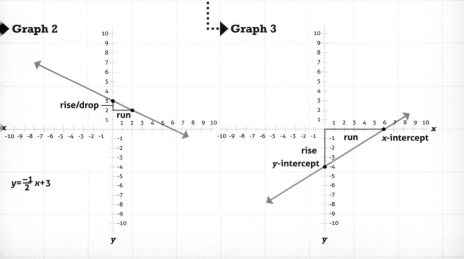

Graph 2 rise/drop $\frac{3}{2}$ run $y = \frac{-1}{2}x + 3$

Graph 3 run x-intercept rise y-intercept

Blaise Pascal

Unlike Descartes, Pascal does not have a town named after him—although there is a street in Paris, and I'm sure many elsewhere, named in his honor. However, he can at least boast of having a unit of measurement named after him: the "pascal" (Pa) is used to measure pressure.

Home Schooling

Blaise Pascal was born in Clermont (now Clermont-Ferrand), France in 1623. His father, Etienne Pascal, a mathematician and scientist, educated his children himself. He wanted Pascal to have a firm grasp of languages so he forbade him from studying mathematics. This, however, only served to pique the young Pascal's curiosity, so he took it upon himself to study mathematics, developing the theorem for angles in a triangle. When Etienne saw this he relented and gave Pascal a copy of Euclid's *Elements* (see pp. 44–45).

While in Paris, Etienne attended meetings of many of the great mathematical minds of France. These meetings were held by a monk, Marin Mersenne, who was also a friend of Descartes. It was at one of these meetings held by Mersenne that Blaise Pascal, still a teenager, handed in his first work on mathematics, "Essay on Conic Sections," published in 1640.

In his early twenties Pascal developed a machine to help his father with his work. Etienne was a tax collector and had to compute many numbers—Pascal's "Pascaline" eased that burden. Also around this time Pascal conducted a series of experiments on pressure, and he used the results of these to argue for the existence of a vacuum. However, when he published his *New Experiment Concerning Vacuums*, in 1647, it led to disputes with other scientists. In fact, Descartes, who we've already met, disagreed with Pascal so much that he wrote that Pascal "has too much vacuum in his head."

In 1653, Pascal wrote his *Treatise on the Equilibrium of Liquid*, in which he gave us what has since become

Pascaline machine In the mid-1600s, Blaise Pascal, along with German astronomer Wilhelm Schickard, invented this mechanical calculator.

known as "Pascal's law of pressure." The law states that if pressure is applied to a non-compressible fluid then the pressure is conveyed to all the fluid and the container. It's something to be thankful for the next time you apply the brakes on your car and the pressure you exert via the brake fluid is transmitted equally to all four wheels.

In the same year Pascal also published *Treatise on the Arithmetical Triangle*. Although Pascal was far from the first to study this triangle, it now has his name attached to it (see pp. 116–21).

Pascal and faith

....

In 1654 a serious accident led Pascal to pledge his life to Christianity. He collected his thoughts on the Christian faith in a work known simply as *Pensées (Thoughts)*, in which, surprisingly, he sets out a logical argument. He states there are four possible consequences to the combination of God's existence or non-existence and your belief or lack of it, as set out in the table below. In short, Pascal said that it's worth believing simply because you stand to lose nothing, but gain everything!

	God exists	God doesn't exist
You believe	You gain everything	You gain nothing
You don't believe	You gain nothing*	You gain nothing

*If God does exist, and He or She is vengeful, then this option could be worse than nothing!

MERSENNE PRIMES

Marin Mersenne (1588-1648) gives his name to a set of primes. These follow the equation: $2^p - 1$, where p is a prime number. This equation does not work for all values of p, however, so it's not automatic that $2^p - 1$ is prime.

Mersenne primes are still being found today. In fact, you can join the Great Internet Mersenne Primes Search (GIMPS)—just download some software and the power of your computer is joined to others around the world. Mersenne primes are largely a curiosity, but large primes do some applications in encryption. Here's a list of some early Mersenne primes:

The prime	Value of $2^p - 1$	Prime/ Not Prime
2	3	Prime
3	7	Prime
5	31	Prime
7	127	Prime
11	2047	Not Prime
13	8191	Prime
17	131071	Prime
19	524287	Prime

Fun with Factorials!

•••

5! What the heck does it mean? Does 5! mean that I'm really excited about the number five? Well, not quite. 5! is actually a factorial, which is effectively 5 • 4 • 3 • 2 • 1. In spite of what you might believe, mathematics is about making your life easier. Multiplication is just a quick way to perform multiple additions, and a factorial (written as *n*!) is a quick way to multiply the natural numbers from one to *n*. Factorials are useful in all sorts of problems, as we'll find out later.

What Is a Factorial?
•••••••••••••••••••••••••••••

The factorial function is defined as $n! = n(n-1)(n-2) \ldots (3)(2)(1)$. Or in plain English $n!$ is the natural numbers from n down to one multiplied together. For example, $5! = 5 \cdot 4 \cdot 3 \cdot 2 \cdot 1 = 120$. The one strange one is $0!$, which equals 1. Working with factorials is quite easy, and most cheap calculators have a factorial button.

Factorials are most often used in probability questions. As an example, you have five guys ready for a police line-up. How many ways can the five men be ordered? To answer this you could try and write out all of the options, but that's unreliable and time consuming. Mathematics is much quicker!

As the police officer sends them in she has choices to make. At first she has five choices—any of the men. For the second person she now has four choices (because one has already

gone in). For the third person she has three choices, then two, then one. All of which means, the number of ways she can arrange the usual suspects in a police line-up is 5!, or $5 \cdot 4 \cdot 3 \cdot 2 \cdot 1$, or 120.

Working with Factorials
•••••••••••••••••••••••••••••••••••

Factorials are fun to work with, mostly because they make you look smart. For example, you might think simplifying $\frac{10!}{9!}$ would be tricky, but the answer is simply 10. I don't even need a calculator to tell you this, because the definition of a factorial pretty much makes the problem go away. Here's a different way of writing the factorial:

$$\frac{10!}{9!} = \frac{10 \cdot 9 \cdot 8 \cdot 7 \cdot 6 \cdot 5 \cdot 4 \cdot 3 \cdot 2 \cdot 1}{9 \cdot 8 \cdot 7 \cdot 6 \cdot 5 \cdot 4 \cdot 3 \cdot 2 \cdot 1}$$

This doesn't look nice, but you can see that most of the numbers in the numerator and denominator can cancel each other out:

There, nice and easy and it makes you look smart!

Now for one last example of onion peeling:

$$\frac{16!}{14! \cdot 5!}$$

To simplify, let's peel the 16! until it looks like a 14!:

$$\frac{16 \cdot 15 \cdot 14!}{14! \cdot 5!}$$

Now we cancel out the 14! above and below the line:

$$\frac{16 \cdot 15}{5!}$$

We can now write out the 5! as a collection of multiplications:

$$\frac{16 \cdot 15}{5 \cdot 4 \cdot 3 \cdot 2 \cdot 1}$$

Basic arithmetic tells us that 5 and 3 can multiply together to give 15, so we can remove them along with the 15 in the numerator to leave:

$$\frac{16}{4 \cdot 2 \cdot 1}$$

Multiply the numbers in the denominator together to give:

$$\frac{16}{8} \text{ or } 2$$

All this work without a calculator. Feel the power of factorials flow through you!

$$\frac{10!}{9!} = \frac{10 \cdot 9 \cdot 8 \cdot 7 \cdot 6 \cdot 5 \cdot 4 \cdot 3 \cdot 2 \cdot 1}{9 \cdot 8 \cdot 7 \cdot 6 \cdot 5 \cdot 4 \cdot 3 \cdot 2 \cdot 1} = 10$$

This approach can be very helpful. For example, if we had $\frac{8!}{6!}$ we could reduce it to $\frac{8 \cdot 7 \cdot 6!}{6!}$ where the 6! on the top and bottom would cancel, leaving $8 \cdot 7 = 56$. I call this "peeling the onion," and it's a great way to imagine working with factorials. Using the above example, the 8! is an onion with eight layers to it while the 6! has six layers. To cancel out the onions you must make them the same size.

So the 8! onion needs to have two layers removed, the eighth layer and the seventh layer. Now in the numerator you have two layers of onion (eight and seven) and an onion with six layers. Now the six-layer onion in the numerator can cancel with the six-layer onion in the denominator.

This is helpful when confronted by a problem like $\frac{100!}{98!}$. This problem cannot be done on a calculator. Just try it; the 100! is too large for a calculator to cope with. Luckily we humans have a brain, so let's use it to peel the onion:

$$\frac{100!}{98!} = \frac{100 \cdot 99 \cdot 98!}{98!} = 100 \cdot 99 = 9900$$

Permutations and Combinations

....

Every January at the school where I teach, students return from two weeks of fun in the snow. Once back in our hallways, there are always a few who forget the combination for their lockers. The funny thing is that what they have forgotten is not a combination; it's really a permutation, and all those locks should really be called permutation locks. Want to know why? Read on.

Permutations

....

A permutation is defined as the number of ways to obtain an ordered subset of r elements from a set of n elements. An example will clear up the jargon.

In the Olympic 100-meters final there are eight people. How many different ways can these eight people stand on the podium (first, second, and third)?

Since we're only looking to put three of the eight people on the podium, these three people are the subset r, while all eight people are represented by n. The fact we're placing them in first, second, and third positions means that they have an order. Therefore we have an ordered subset of three people from a set of eight people.

The notation for permutations looks like $_nP_r$ or $P(n, r)$, where n is the total number of objects and r is the number of arranged objects. In fact, most cheap scientific calculators have an $_nP_r$ button.

The formula that we should use to find our answer is:

$$_nP_r = \frac{n!}{(n-r)!}$$

For the above example we'd have:

$$\frac{8!}{(8-3)!} = \frac{8!}{5!} = \frac{8 \cdot 7 \cdot 6 \cdot 5!}{5!} = 8 \cdot 7 \cdot 6 = 336$$

1

2

3

Another way to look at this problem is to approach it like the police line-up on the previous page. How many choices do we have for the winning runner? Any of the runners could win, so we have eight. For the second-place runner we have seven choices (because one has already crossed the line) and for the third-place runner we have six. Multiplying these together gives us the same answer, 336.

Combinations

A combination is very similar to a permutation, but there is one significant difference. Whereas a permutation is the number of ways to obtain an ordered subset of r elements from a set of n elements, a combination is the number of ways to obtain an unordered subset of r elements from a set of n elements. The notation for combination is $_nC_r$ or $\binom{n}{r}$. The formula we use is:

$$_nC_r = \frac{n!}{(n-r)!r!}$$

Let's look at the following example. In the heats for the Olympic hundred meters, the first three across the line advance to the next round. In a field of eight competitors, how many ways can three competitors advance? It no longer matters if you're first, second, or third. This makes the subset r an unordered subset. To calculate we get:

This means that while there were 336 different permutations for who came home with the medals from the final, there are just 56 combinations for the different athletes that can make it through the qualifying heats.

The sharp-eyed among you might have noticed that the number of combinations is less than the number of permutations. The value of a combination will always be less than or equal to the value of the permutation; in fact, there's a formula for the relationship between permutations and combinations, which is:

$$_nC_r = \frac{_nP_r}{r!}$$

To Order or Not to Order?

Often people forget which function to use when the order is important and when the order is unimportant. I just use a simple mnemonic to remind myself: permutations are picky about position, while combinations couldn't care less.

Now let's go back to our school lockers. Why should our combination locks really be called permutation locks? Well, a lock requires the student to input a few numbers in a certain order, let's say three numbers. A lock that is set to 33, 21, 45 will not open if you input 21, 33, 45. Therefore the order is important and that lock should really be called a permutation lock.

$$_nC_r = \frac{8!}{(8-3)!3!} = \frac{8!}{5! \cdot 3!} = \frac{8 \cdot 7 \cdot 6 \cdot 5!}{5! \cdot 3!} = \frac{8 \cdot 7 \cdot 6}{3 \cdot 2 \cdot 1} = 56$$

Pascal's Triangle

Pascal's triangle is full of mathematical goodness—a whole world of cool calculations is contained within. But it should be pointed out that Pascal was not the first mathematician to discover this triangle. References to what we call Pascal's triangle can be found in Chinese, Indian, and Persian mathematics well before the birth of Pascal. It's just the Western bias we've seen before that attaches his name to it.

Making Pascal's Triangle

Pascal's triangle goes back a long way. Its connection to the expansion of binomials (see pp. 122–23) made it very useful in early mathematics, and early references to something almost exactly like it can be found as early as the work of sixth-century Indian mathematician Varahamhira.

Constructing Pascal's triangle is simple. You start with one at the apex then diagonally below it you write two more ones. As you descend through the rows, the ones continue along the left and right diagonals, and the numbers inside the triangle are found by adding the two numbers that are immediately above left and above right.

Patterns in Pascal's Triangle

Inside Pascal's triangle is a wonder of mathematical patterns. The first left and right diagonals all contain the number one. In the second diagonals you can find the natural number set; the third diagonals contain triangular numbers, and every second number in the third diagonal is a hexagonal number; while the fourth diagonals contain tetrahedral numbers. There are patterns of other exotic-sounding numbers such as pentatopes (strange four-dimensional tetrahedrons) and Catalan numbers, but unfortunately we don't have space for them here.

Building Pascal's triangle is quite simple.

Powers in Pascal's Triangle

....

Another interesting bit of mathematics hidden in Pascal's triangle can be found if you sum the numbers of its rows. The sum of the first row is one; the second row gives two; and the third row gives four. The fourth row sums to eight, the fifth gives sixteen, and so on. You may have noticed that the sum of the digits in each row of Pascal's triangle is a power of two (see top right).

Row in Pascal's Triangle	Sum of Numbers	Power of two
1	1	2^0
2	2	2^1
3	4	2^2
4	8	2^3
5	16	2^4
6	32	2^5

Along with powers of two, Pascal's triangle also contains powers of eleven. If you look at the first row, you have the number one. This is 11^0 (anything to the power zero is one). The second row can be read as 11, which is 11^1, while the next row can be viewed as 121 or 11^2. From this start it's easy to guess the value of 11^3, it's 1331; and then 11^4 is 14641.

At this point it gets a little more complicated: 11^5 is 161051—not the numbers of the fifth row of Pascal's triangle. The reason for this is we are now faced with two-digit numbers. However, we can still make the pattern work because the digits in the triangle represent powers of ten. Starting from the right we have the units, then the tens, hundreds, thousands, and so on. In the fifth row the hundreds and thousands spots both have a 10 in them. This means we have to carry forward the hundreds to the thousands column, and the thousands to the tens-of-thousands column.

Let's take a closer look. The numbers of the fifth row, where we should find 11^5, are 1, 5, 10, 10, 5, 1. The 1 on the right represents the units, the 5 is in the tens column, and so on. This can be shown as:

fifth row:	1	5	10	10	5	1
1s						1
10s					5	0
100s				10	0	0
1000s			10	0	0	0
10,000s		5	0	0	0	0
100,000s	1	0	0	0	0	0

Carrying over the 10s in the hundreds and thousands columns gives 0 in the hundreds column, 1 in the thousands column, and 6 in the tens of thousands column. So:

$11^5 =$	1	6	1	0	5	1

Therefore, with a little carrying forward you have all the powers of eleven.

Pascal's Triangle Part 2

As we've already discovered, there is a sense of beauty and symmetry to Pascal's triangle; and like that of the Fibonacci sequence and the golden ratio, it's a beauty that requires no great understanding of higher mathematics.

The Beauty of Pascal's Triangle

To unlock the beauty in Pascal's triangle all you need is a set of crayons and a sense of adventure. Choose a set of numbers with something in common—for example, multiples of five—and color them in. Alternatively, pick any number and divide the numbers in Pascal's triangle by it, but leave remainders rather than using decimal places. The remainders will range from zero to one less than the number you chose—assign colors to different remainders and color in those spots. If you do either of the above you'll start to see some amazing patterns emerge; use your imagination and start searching for other patterns.

Using the previous approach to obtain remainders, but dividing by two and coloring a remainder of one while leaving a remainder of zero

Sierpinski's triangle—by continually dividing an equilateral triangle into further congruent triangles, you get an ever more wonderful fractal pattern.

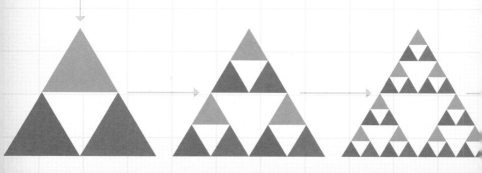

white, will form what is known as Sierpinski's triangle. The triangle is named after the Polish mathematician Waclaw Sierpinski (1882–1969), who described it in 1915. This shows that even in the modern era we are still making new connections to mathematics that has been known for centuries.

This is an example of the triangle's connection to fractal geometry (you know, the type of thing that shows up on those posters of weird spirals and patterns). A relatively new branch of mathematics, the term "fractal" was only coined in 1975 by French mathematician Benoit Mandelbrot.

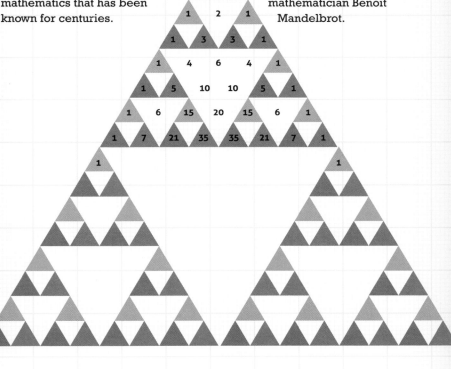

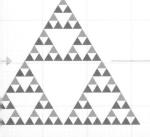
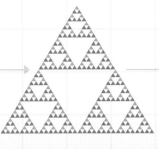
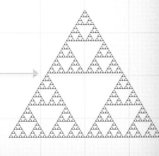

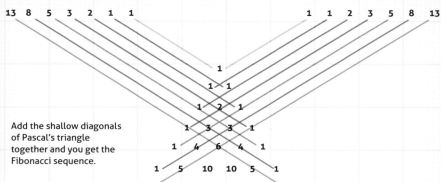

Add the shallow diagonals of Pascal's triangle together and you get the Fibonacci sequence.

Pascal's Triangle and Fibonacci

Another neat trick is to take the shallow diagonals of the triangle and add the terms together (see above) to get the Fibonacci sequence. This connects three of the coolest things in mathematics: Pascal's triangle, the Fibonacci sequence, and the golden ratio.

Pascal's Hockey Sticks

For me, it's nice to see Canada's national sport reflected in Pascal's triangle. If you add the numbers in any diagonal you will find the sum of those numbers in the next row. For example, let's add a few of the terms in the fifth diagonal on the left. In this diagonal the numbers are 1, 5, 15, 35, 70, and so on. If we sum these we get 126—

one row down and one spot to the left. So the numbers form the shaft and the sum forms the stick's blade. On the right let's look at the third diagonal and add the first four numbers: 1, 3, 6, 10. Their sum is one row down and one spot to the right. The shaft of the stick can be any length but must start with a 1; the blade of the stick is always one down and one to the left or right, depending on where you start.

Pascal's Triangle and Flowers

One last neat property of Pascal's triangle is Pascal's petals. If you take any term, other than the ones on the edge, it will be surrounded by six other terms like the petals of a flower. If you multiply the three terms opposite each other you will find they have the same product as the other three terms.

One last neat property of Pascal's triangle is Pascal's petals ... If you multiply the three terms opposite each other you will find they have the same product as the other three terms.

As an example, let's look at the term six rows down and three to the right. This has a value of 10. Around it we have the numbers 4, 6, 10, 20, 15, and 5. If we multiply the non-adjacent numbers 4, 10, and 15 we get 600. If we multiply the remaining non-adjacent numbers 6, 20, and 5 we get 600. Pretty neat, right?

The Binomial Theorem

The binomial theorem is a nice simple way to expand (or multiply out) binomials. The biggest application of the binomial theorem is within probability problems in which there are only two choices, to you and me that's something like tossing a coin. This might seem limiting, but in reality it's very easy and useful to divide things into two groups.

Heads or tails?

Let's Expand Some Binomials

If we expand $(x + y)^0$ we get one, as anything to the zero power is one.
If we expand $(x + y)^1$ we get $1x + 1y$.

Then, if we expand $(x + y)^2$ we need to write it out twice $(x + y)(x + y)$ and "foil". This gives $1x^2 + 1xy + 1xy + 1y^2$, which when you collect like terms is $1x^2 + 2xy + 1y^2$.

If we expand $(x + y)^3$, we need to write it out three times as:

$$(x + y)(x + y)(x + y)$$

Next we "foil" out the first two binomials and simplify. This gives:

$$(x^2 + 2xy + 1y^2)(x + y)$$

Then the trinomial and the binomial need to be multiplied, giving:

$$1x^3 + 1x^2y + 2x^2y + 2xy^2 + 1xy^2 + 1y^3$$

After this we need to collect like terms to give:

$$1x^3 + 3x^2y + 3xy^2 + 1y^3$$

As you can imagine, this gets old very quickly. I love mathematics, but even I don't feel like writing out the next expansion, so I won't. Luckily there's a nicer way to deal with binomial expansions. Here's what we have so far:

$$(x + y)^0 = 1$$
$$(x + y)^1 = 1x + 1y$$
$$(x + y)^2 = 1x^2 + 2xy + 1y^2$$
$$(x + y)^3 = 1x^3 + 3x^2y + 3xy^2 + 1y^3$$

If we look at the coefficients of the expansions, we see they are numbers from Pascal's triangle. If we wanted to expand the binomial $(x + y)^4$, we could do it without all the mess presented above. Here's $(x + y)^4$ expanded:

$$(x + y)^4 = 1x^4 + 4x^3y + 6x^2y^2 + 4xy^3 + 1y$$

Now the coefficients are just one aspect of the expansion, another is the variables. If you look at the last expansion, the x variable starts with an exponent of four and decreases until the x variable disappears on the last term. (In fact, it's there but is x^0, which as we've said before is equal to one.)

So the exponents on x go: four, three, two, one, and zero. The y variable starts with an exponent of zero and rises to an exponent of four. Also, the sum of exponents on any term in this expansion is four. The focus on four is due to the exponent of the original binomial, which is $(x + y)^4$. So if we wanted to expand $(x + y)^5$, we could just follow Pascal's triangle and add in the variables:

$$(x + y)^5 = 1x^5 + 5x^4y + 10x^3y^2 + 10x^2y^3 + 5xy^4 + 1y^5$$

Let's Expand with Factorials
••••••••••••••

What happens if you are asked to expand $(x + y)^{13}$? Do you really want to write out fourteen rows of Pascal's triangle just to get the coefficients to the expansion? No! Luckily you can use factorials to find the coefficients of the expansion. Looking at the last example:

$$x^5 + 5x^4y + 10x^3y^2 + 10x^2y^3 + 5xy^4 + 1y^5$$

The coefficient of the second term is five. This number can be found by writing a factorial statement. The sum of exponents on x and y is five, x has an exponent of four, and y has an

exponent of one. We can write this as $\frac{5!}{4! \cdot 1!}$ which equals five. The next term has three for the x exponent and two for the y exponent. This can be written as $\frac{5!}{3! \cdot 2!}$ which equals ten. So the coefficient is:

$$\frac{\text{(sum of exponents)!}}{\text{(first exponent)!(second exponent)!}}$$

Now, if we wanted to expand $(x + y)^{13}$—and why wouldn't we?—we could write out the variables first then add the coefficients later. The first five terms without coefficient would look like:

$$x^{13} + x^{12}y + x^{11}y^2 + \dots$$

Adding the coefficients gives:

$$\frac{13!}{13!0!}x^{13} + \frac{13!}{12!1!}x^{12}y + \frac{13!}{11!2!}x^{11}y^2 + \dots$$

Evaluating the factorials gives:

$$1x^{13} + 13x^{12}y + 78x^{11}y^2 + \dots$$

Another Way to Expand
••••••••••••••••••••••••••

Another way to find the coefficients of a binomial expansion is to use combinations. Using the last example of $(x + y)^{13}$ we could expand the first few terms as:

$$_{13}C_0 x^{13} + {}_{13}C_1 x^{12}y + {}_{13}C_2 x^{11}y^2 + \dots$$

All three methods are essentially the same, just use whatever suits you.

Leonhard Euler

One of the most prolific mathematicians ever, Leonhard Euler, was born on April 15, 1707. His father Paul had some mathematical training and taught his son elementary mathematics. Leonhard entered university in 1720 and his potential was soon discovered. In 1723 Euler received a master's degree in philosophy.

St. Petersburg, Berlin, and Back

In 1726, when Nicolaus Bernoulli, the famous Johann Bernoulli's eldest son, died Euler replaced him at the Imperial Russian Academy of Science in St. Petersburg. Euler lived and worked with Bernoulli's second son, Daniel. However, Daniel became disenchanted with the Academy, and in 1733 he left, leaving Euler to replace him as senior chair of mathematics. In 1734 Euler married and eventually had thirteen children, only five of who survived infancy.

In 1741 Euler took a position in Berlin, where he would spend the next twenty-five years. During that time he wrote many articles and in 1759 he assumed leadership of the Berlin Academy. In 1766, Euler returned to the Imperial Russian Academy, but very soon afterwards he became blind. In spite of losing his sight, Euler continued to work with the help of his sons Johann and Christoph. In fact, Euler continued to publish until he passed away in St. Petersburg on September 18, 1783.

The Bridges of Königsberg

The "Bridges of Königsberg" is a classic problem. The town, then part of Prussia, but now Kalingrad, Russia, is on a river with two islands in it. There are seven bridges connecting the island to both sides of the river and to each other. The question was whether there was a way to start in one location and cross each bridge once and only once.

It's a strange task, to be sure, and one that Euler proved was impossible. The reason is based on the number of bridges to each land mass. For the land masses you don't start at or finish at, you must enter and leave. This requires two bridges, which

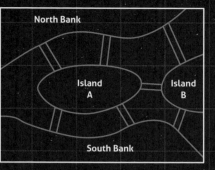

The Bridges of Königsberg can be reduced to nodes (land masses) and lines (bridges). Many maps follow this format; for example, train maps often just show the stations as nodes and the track as lines. Euler's path also has practical applications; for example, when haulage companies are looking to reduce fuel and travel costs and want to plan routes that do not involve doubling back on themselves.

means that any land mass that is not the start or finish point must have an even number of bridges. This is not the case for Königsberg, where all four landmasses have an odd number of bridges. This leads to a thing called an "Eulerian path," which will allow you to cross each bridge (or edge) only once, if and only if two of the land masses (or nodes) has an odd number of bridges. This would be possible if we removed the bridge between the two islands. An Eulerian cycle requires you to return to your starting point and it also requires that all of the land masses (nodes) should be attached to an even number of bridges (edges).

Related to the "Bridges of Königsberg" question, Euler also gave us the formula $V - E + F = 2$, where the V is the number of vertices, E the number of edges, and F the number of faces of a polyhedron.

A NOTE ON NOTATION

Euler is also responsible for producing some of the notations we use today. He gave us $f(x)$ for a function applied to a variable x, the Greek letter Σ (sigma) for summation, the letter i for imaginary numbers, and e for the value 2.71828... The first reference to e comes from a work published by John Napier (see p. 136), but the discovery of the constant is credited to Jacob Bernoulli, the son of Euler's mentor. Euler started using the letter e for this constant and it stuck. The number e is another fabulous number, like π and φ (see pp. 20-21 and 88-89). The number e can be calculated using an infinite series as follows:

$$e = \frac{1}{0!} + \frac{1}{1!} + \frac{1}{2!} + \frac{1}{3!} + \frac{1}{4!} \cdots$$

This leads to one of the most beautiful equations in mathematics: $e^{i\pi} + 1 = 0$. I have thought of putting this on a T-shirt but I don't think people will like it as much as my π one.

Hit the Road
. . . .

It takes a certain kind of person to be a long-distance runner. They have strength, stamina, resolve, and a zen-like ability to say no to friends on a Friday night, opting instead to rest up in prep for the early-Saturday-morning park run. And they're obsessive too, always finding ways of pushing themselves; it's the long-distance runner you see jogging to school every morning through hail, sleet, and snow, their lycra gear soaked through, taking circuitous routes just to extend the run by a few more minutes.

The Long Run
. . . .

In honor of these extraordinary beings, here is a Euler-inspired challenge to revise what was covered on the previous page. The map (right) details a running route that passes through seven points, A to G. Is it possible to run the complete course, starting and ending at A and running all the roads once? If not, which changes can be made to the route to make this possible? Is it possible to cover the entire course without running along any road twice? Have a go at solving it first, before reading on.

Maybe the jogging routes you devise are much simpler, but can you solve this Euler-inspired puzzle?

An Eulergy

• • • • • • • • • • • • • • •

This is of course a classic Euler-cycle problem. Looking at the map, there are seven "nodes," or more simply seven places where roads meet. There are only two roads leading to and from A, so this is an even node. The other six nodes, however, are all odd. Therefore, in its present configuration there is no Euler cycle.

In order to have an Euler cycle we need the number of sections joining each node to be even. If we run a section twice it's like adding another road. In the spirit of the long-distance runner, let's opt for the longer roads, removing the shortest.

By running from A to B and then to C, we can remove road 2 and make node C even. Then we can run from C to D and then to E and remove road 6, making nodes D and G even. We can then run from E to F to G and back to F. If we run over road section 9 again, it's like adding another road, which makes F and E even. We can then run from E to B. Repeating the road section 1 again will make B and A even.

This will then be an Euler cycle.

With the removal of two short sections and the repetition of two sections an Euler cycle can be made. The route would follow the road sections 1 → 3 → 4 → 5 → 9 → 7 → 8 → 9 → 10 → 1.

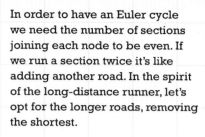

E

10

B

3

1

This is of course a classic Euler-cycle problem. Looking at the map, there are seven "nodes," or more simply seven places where roads meet.

Carl Friedrich Gauss

Have you ever met someone who's a real know-it-all? When you tell them something you have just learned, they reply: "Yeah, I knew that." Sometimes you believe them, but most of the time you think they're just full of it. Well, Carl Friedrich Gauss was that sort of guy, but he really did know it all. When we talk about freaky geniuses, Carl Friedrich Gauss is our best example.

Early Life

Gauss was born in Brunswick, Germany, in 1777. At the age of seven, in elementary school, he impressed his teacher with his brilliance. Aged eleven he entered secondary school learning languages. In 1792, at fifteen, he entered Brunswick Collegium. When there Gauss independently discovered the binomial theorem (see pp. 122–23). Three years later Gauss left for Göttingen University. However, he left Göttingen without a degree and returned to Brunswick, receiving a degree in 1799. Gauss had been receiving a stipend from the Duke of Brunswick and, at his request, Gauss submitted a doctoral dissertation to the University of Helmstedt. Gauss's dissertation was on the fundamental theorem of algebra.

A Busy Few Years

The Duke of Brunswick, Gauss's supporter, died and in 1807 Gauss took the position of director of the Göttingen Observatory. Previously Gauss had accurately predicted the position of two new celestial objects, Ceres (now classified as a dwarf planet) and Pallas (now classified as an asteroid). In 1808 Gauss's father died. Then, in 1809, Gauss's wife Johanna died during the birth of their second son, Louis, who only lived until 1810. A year later Gauss married Johanna's best friend Minna. In the years from 1805 to 1811, Gauss married twice, fathered four children and suffered three bereavements. In anyone's book, that's a lot to deal with.

"It may be true that men who are mere mathematicians have certain shortcomings, but that is not the fault of mathematics, for it is true of every other occupation."

—Carl Friedrich Gauss

DISQUISITIONES ARITHMETICAE: Other than his dissertation, the *Disquisitiones Arithmeticae*, in 1801, was Gauss's first published work. It dealt mostly with number theory which includes the study of primes (see p. 16) and Diophantine equations (see pp. 54-55).

THEORY OF CELESTIAL MOVEMENT: In 1809 Gauss published his second work. The book flowed naturally from his predictions of the paths of Ceres and Pallas—a dwarf planet and asteroid that were at the time thought to be new planets.

ARTICLES: Gauss also published many articles including papers on series, integration, statistics, and geometry.

Later Years

In 1831 Gauss began to work with Wilhelm Weber (1804–91), a German physicist who had come to Göttingen at Gauss's request. Together they produced many papers until Weber had to leave Göttingen in 1837.

Gauss's work lives on in many forms. Two examples are that the SI unit of magnetic flux bears Weber's name; and the process of "degaussing," which removes the magnetic field from an object, is also named for Gauss. It's this that makes that weird sound when you switch your television or computer monitor on.

The Fundamentals of Algebra

One of Gauss's most important contributions to mathematics was his work into the "fundamental theorem of algebra." The theorem states that a polynomial with real or complex coefficients will have root(s) in the complex plane. Plainly stated, if you have a polynomial with a degree n, and the coefficients are real or complex numbers then you will have n roots or solutions. For example, on pages 102–3 we looked at quadratics (parabolas) which had a degree of two and in all three cases there were two solutions—two unequal real solutions, two equal real solutions, and two complex solutions.

By the age of thirty Gauss had been appointed director of Göttingen Observatory.

6

Talking
in Code

As we've strolled through the history of algebra we've found many situations in which algebra is put to good use. Here, as well as looking at logarithms, let's visit an area in which algebra plays a big role in our everyday lives: privacy. We haven't the space to dissect the labyrinthine encryptions of today's computers, so instead we'll visit the more accessible but equally fascinating subject of basic codes.

Exponents Laws

Before we get stuck into money, and interest rates in particular, we should have a quick recap of exponents. An exponent simply represents successive multiplications; that is to say, 2^5 (the ^{5} is the exponent) is the same as $2 \cdot 2 \cdot 2 \cdot 2 \cdot 2$.

Power Play

A power has three parts: a coefficient, a base, and an exponent. For example, in $3x^5$, 3 is the coefficient, x is the base, and ^{5} is the exponent. Let's have a look at the laws governing exponents.

1) $x^n \cdot x^m = x^{n+m}$

When you multiply powers that have the same base, you simply add the exponents together. For example:

$x^2 \cdot x^3 = x^5$ or $(x \cdot x) \cdot (x \cdot x \cdot x) = (x \cdot x \cdot x \cdot x \cdot x)$

2) $x^n \div x^m = x^{n-m}$

When you divide powers that have the same base you subtract their exponents. For example:

$x^7 \div x^4 = x^3$ or
$\frac{(x \cdot x \cdot x \cdot x \cdot x \cdot x \cdot x)}{(x \cdot x \cdot x \cdot x)} = \frac{(x \cdot x \cdot x \cdot \cancel{x \cdot x \cdot x \cdot x})}{(\cancel{x \cdot x \cdot x \cdot x})} = (x \cdot x \cdot x) = x^3$

3) $(x^n)^m = x^{n \cdot m}$

When you raise a power to an exponent, multiply the exponents. For example:

$(x^3)^2 = x^6$ or
$(x^3) \cdot (x^3) = x^6$ (using law 1)

4) $(xy)^m = x^m y^m$

In other words, when you have two or more bases raised to the same exponent you can bring the exponent into each base. For example:

$(xy)^3 = (xy)(xy)(xy) = (x \cdot x \cdot x)(y \cdot y \cdot y) = x^3 y^3$

5) $\left(\frac{x}{y}\right)^m = \frac{x^m}{y^m}$

In other words, when you have a fraction raised to an exponent, the exponent can be brought into the numerator and the denominator. For example:

$\left(\frac{x}{y}\right)^3 = \left(\frac{x}{y}\right)\left(\frac{x}{y}\right)\left(\frac{x}{y}\right) = \frac{(x \cdot x \cdot x)}{(y \cdot y \cdot y)} = \frac{x^3}{y^3}$

6) $x^0 = 1$ where $x \neq 0$

An example by expansion:

$\frac{x^3}{x^3} = \frac{(x \cdot x \cdot x)}{(x \cdot x \cdot x)} = \frac{(\cancel{x \cdot x \cdot x})}{(\cancel{x \cdot x \cdot x})} = 1$

and by using law two $\frac{x^3}{x^3} = x^{3-3} = x^0$, since mathematics must maintain consistency then $x^0 = 1$, therefore anything to the zero power is one, except 0^0.

7) $x^{-m} = \frac{1}{x^m}$

This relates back to the law six and the reverse of law two, for example:

$x^{-4} = \frac{1}{x^4}$ because

$x^{-4} = x^{0-4} = \frac{x^0}{x^4} = \frac{1}{x^4}$

8) $\frac{1}{x^{-m}} = x^m$

Relates back to laws six and two, for example:

$\frac{1}{x^4} = x^4$ because

$\frac{1}{x^4} = \frac{x^0}{x^4} = x^{0-(-4)} = x^{0+4} = x^4$

Note that for laws seven and eight it's easier just to remember that you change the sign on the exponents when you change the side of the fraction; for example:

$\frac{2^{-3}}{3^{-2}} = \frac{3^2}{2^3} = \frac{9}{8} = 1.125$

You may notice that the introduction of these exponent laws follows the order of number sets introduced on pages 14–5. The first five exponent laws dealt with natural numbers; law six introduced a zero exponent thus introducing the whole number set; and laws seven and eight introduced the integer set to exponents. Now we will add two more laws—actually two parts of one law, but it's easier if we separate

it into two—that will expand our laws to include rational numbers, or fractions.

9) $\sqrt[n]{x} = x^{\frac{1}{n}}$

For example, lets find the square root of x. Firstly the square root should be written as $\sqrt[2]{x}$, which would make it equal to $x^{\frac{1}{2}}$. The index on the square root is assumed but you see it on cube roots and up. So if we multiply the square root of x by itself we get:

$\sqrt{x} \cdot \sqrt{x} = \sqrt{x} \cdot x = \sqrt{x^2} = x$

If we multiplied the fractional exponent by itself, we'd get the same result. (There are some restrictions, see box.) Therefore $3\sqrt{x}$ would equal x and so on. The last law is just an extension of law nine.

10) $\sqrt[n]{x^m} = x^{\frac{m}{n}}$ or $(\sqrt[n]{x})^m = x^{\frac{m}{n}}$

For example, $\sqrt[3]{x^2} = x^{\frac{2}{3}}$ or $(\sqrt[3]{x})^2 = x^{\frac{2}{3}}$

CAUTION REQUIRED

When dealing with exponents, we have to be careful because they can sometimes generate unusual results. For example, let's take the square root of minus two squared: $\sqrt{-2^2}$, following BEDMAS we square the -2 first to get 4, then the square root of 4 is 2. When you take $(\sqrt{(-2)})^2$, the square root is done first to give $\sqrt{2} \cdot i$, which we square to give $2 \cdot i2$, which results in an answer of -2. As you can see the order has a big impact on the result.

Exponential Equations

Exponents are used in finance (compound interest), biology (growth and decay), physics (radioactivity), chemistry (reaction rates), economics (supply and demand curves), and elsewhere. In my home province of British Columbia, the rate at which the pine forests are dying—due to the mountain pine beetle—is exponential.

An exponential equation is an equation in which the variable is in the exponent. This should not be confused with any equation with an exponent. So, for example, $2^x = 8$ is an exponential equation where $x^2 = 9$ is not. However, before we take a look at a couple of real-world examples of exponential equations, let's look at how to solve them.

Solving Exponential Equations

Solving exponential equations can be very straightforward. An example: when seeing $2^x = 8$ many people will say $x = 3$, which is correct. This kind of intuitive feel for numbers is hard to explain. What happens is that you change the 8 into a power of 2 so that, $2^x = 2^3$ then, because the bases are the same you can compare the exponents and you find that $x = 3$.

Another example, $3^{2x-1} = 27$, can be solved the same way. Because 27 is equal to 3^3 we can write the equation

as $3^{2x-1} = 27$. Now that the bases are the same we can compare the exponents and get $2x - 1 = 3$. To solve for x we add 1 to both sides so that we get $2x = 4$. Dividing both sides by 2 we finish with $x = 2$.

With this next example, $2 \cdot 3^x = 162$, there is the desire to multiply the 2 and 3 at the start. But only the 3 is "holding up" the x, not the 2. So we must isolate the exponential 3^x by dividing both sides by 2, giving us $3^x = 81$. Then 81 is equal to 3^4; therefore $3^x = 3^4$ and finally $x = 4$. Which is all very nice, but some simple-looking equations, such as $2^x = 12$, cannot be solved in this manner. All we can do is say that the answer is between 3 and 4 ($2^3 = 8$ and $2^4 = 16$). To find a more exact solution requires logarithms (see pp. 136–37).

The Mountain Pine Beetle

The mountain pine beetle is a big problem in British Columbia. Since the late 1990s, these little guys have been tearing though the forests of the interior of the province. The growth in the infested area is an exponential equation. Using a technique called regression, which we won't go into, an equation can be found that best matches the data in the table (right). The equation it gives is $A = 63(2.32)^t$, where A is the area infested and t is the number of years since 1998 (for 1999, $t = 1$); this allows foresters to predict the level of devastation. In practice the equation begins to fail as the food supply decreases. What happens is the pine beetles continue to spread until they run out of trees to munch on, when they die off.

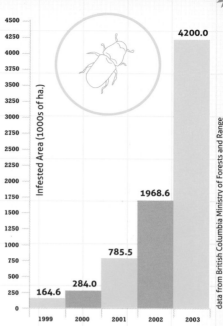

Infested Area (1000s of ha.)

4200.0

1968.6

785.5

284.0

164.6

1999   2000   2001   2002   2003

data from British Columbia Ministry of Forests and Range

Radioactive Decay

Another example of the importance of exponential equations can also be found in Canada. As it turns out, Canada produces more than half of all medical isotopes (radioactive chemicals) in the world. In December 2007 a reactor producing the isotopes shut down, creating a global shortage. Many people complained that the laboratory where the isotopes are produced should have created a stockpile. However, this exposes a common misunderstanding: When most people think about radioactivity, they think of long-term radioactivity because of "The Bomb." This is scary stuff, but many radioactive elements have short half-lives (the time it takes for half the matter to decay); for example, iodine-131, used in the treatment of thyroid cancer, has a half-life of just eight days.

So to stockpile iodine-131 the facility would have to compensate for radioactive decay by producing more than was needed to start with. For argument's sake, let's say that we needed 100 kg at the end of a shutdown lasting thirty-two days. The equation for a chemical's half-life is $F = I(\frac{1}{2})^{\frac{d}{8}}$, where F is the final amount, I is the initial amount, and d is the number of days. We know $F = 100$ and $d = 32$, so we can solve for I.

$$100 = I(\tfrac{1}{2})^{\frac{32}{8}}$$
$$100 = I(\tfrac{1}{2})^4$$
$$100 = I(\tfrac{1}{16})$$
$$1600 = I$$

Therefore the laboratory would have to produce sixteen times the amount needed at the start of shutdown to have enough left at the end of the month.

Logarithms

Now we've met exponents it's time to meet their evil twin: logarithms. Mathematics is essentially about going somewhere and then getting back. Most of the time we learn about how to do something, then we learn how to reverse it. First we learn to add, then subtract. We learn to multiply, then divide. We learn about squares, then square roots. Exponents and logarithms are exactly the same, and a logarithm is simply the inverse of an exponent.

Logarithms: The New Kids in Town

Logarithms are a relatively new addition to mathematics. The first published account of logarithms was in 1614 by the Scottish mathematician John Napier (1550–1617), in his book *Mirifici Logarithmorum Canonis*. Around the same time, the Swiss mathematician, Joost Bürgi (1552–1632), discovered logarithms independently but did not publish his findings until four years after Napier.

Logarithms were initially developed to help with difficult multiplication and division problems, but since the advent of the calculator and computer this application has become pretty much redundant. Logarithms were the basis of the slide rule—a device carried around by every science and math geek in the '50s and '60s. However, by the '80s almost all of them had switched to the calculator—though they,

myself among them, were still geeks! Logarithms still have uses today, mostly to help compare magnitudes of numbers. The most common logarithm uses a base of ten.

The function of a logarithm or "log" is to give the exponent on a base of ten that would equal a particular number. For example, $\log(10)$ is equal to 1. This is because $10 = 10^1$. The $\log(100) = 2$ because $100 = 10^2$. The $\log(1000) = 3$, the $\log(10000) = 4$, and so on. The $\log(250) \approx 2.4$ because $250 \approx 10^{2.4}$. This has the effect of taking a wide range of numbers and reducing them to smaller, more-manageable numbers. In fact, all the numbers from one to a billion can be written as 0 to 9.

Practical Uses of Logarithms

One practical use of logarithms is the Richter scale, which measures the magnitude of an earthquake. Because it's a logarithmic scale, a change of one on the Richter scale corresponds to a tenfold increase in magnitude. So, if one earthquake measures four while another measures seven, the difference in magnitude is not three but a thousand ($7 - 4 = 3$; $10^3 = 1000$). This is why an earthquake measuring four is hardly noticed while one measuring seven is very significant.

The pH scale measures how acidic or basic something is. It works in much the same way, but is the negative log of the hydrogen-ion concentration, therefore things that are more acidic have smaller numbers. The pH scale goes from 1 to 14, where 1 is the most acidic and 14 is the least acidic (most basic). As an example, let's say milk has a pH of 6.5 and soda has one of 2.5. This would make the soda 10,000 times more acidic ($6.5 - 2.5 = 4$; $10^4 = 10,000$) or conversely the milk 10,000 times more basic (generally speaking).

The decibel (dB) scale differs from the Richter scale in one key aspect. For each ten points on the decibel scale there is a tenfold increase in sound intensity. The simplest way to work with the decibel scale is to divide the numbers by ten and treat it like the Richter scale. For example, the difference between loud music at 100 dB and normal conversation at 60 dB is a difference in intensity of 10,000. To follow this calculation, divide both dB measures by ten, then subtract one from the other to find the difference, which is four. The difference in intensity is 10^4 or 10,000.

Hearing damage due to short-term exposure starts at 120 dB, and hearing damage due to long-term exposure (greater than eight hours) at 85 dB. For every 5 dB increase in intensity the exposure time decreases by half. Therefore, if you're being blasted with 90 dB you can last four hours before long-term damage. At 95 dB it's two hours, 100 dB is an hour, and so on. A level of 120 dB causes long-term damage after just four minutes.

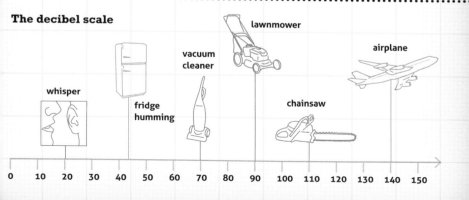

The decibel scale

whisper · fridge humming · vacuum cleaner · lawnmower · chainsaw · airplane

0 10 20 30 40 50 60 70 80 90 100 110 120 130 140 150

Codes and Ciphers

· ·

**Moving on from money, we reach the related field
of privacy. After all, where would we be without
password-protected accounts, personal identification
numbers, and the like? The ability to send and receive
messages without them being read is very important;
and the art and mathematics of cryptography, which
includes codes and ciphers, has been around since
the time of the ancient Greeks.**

Encrypted World

· ·

Throughout most of history, only the
most secret of political, military, or
economic secrets were sent in code
or cipher, and only a select few were
involved with cryptography.
However, in the modern age almost
everyone in the developed world
deals with cryptography in their
daily lives. Every time you use your
credit or debit card, every time you
use a "secure" website, every time
you unlock your car remotely, you're
using cryptography. Yet many people
remain unaware of its significance.

Codes

· · · · · · · · · ·

Codes and ciphers are actually
different things, though many people
use the words interchangeably.
Strictly speaking, a code is a secret
language in which a word is given
meanings in place of its normal ones.
It is the kind of thing that you hear in
classic spy movies, and that was
broadcast to the French Resistance
during the Second World War in the

form of such oblique messages as "Les
carottes sont cuites" ("The carrots are
cooked"). The American forces also
employed code talkers, most famously
Navajos, who adapted their language
to send secret military messages.

Ciphers

· · · · · · · · · · · ·

Unlike codes, ciphers are not a secret
language, but rather a way to change
language so it's unintelligible to those
without a key. A key is a system to
encipher and decipher messages.
A sender will encipher "plain text"
to create a "ciphered text"; while the
receiver will use the same key to
decipher the "ciphered text" back into
"plain text." Whichever method is
used, good encryption requires that
a lot of work is required to find or
crack the key.

Solving Ciphers

· ·

To conclude our journey through
algebra we'll take a closer look at two
enjoyable ciphers: the simple Caesar
cipher and the slightly more complex

Vigenère cipher (see pp. 140–41), which is an advancement on the Caesar cipher, and was still in use during the American Civil War. Both of them are substitution ciphers, in which letters are substituted for one another. It's a form of ciphering that you may already be familiar with from the cryptograms that appear in Sunday newspapers.

These puzzles can be challenging, but their solution only requires the application of an algorithm and a degree of patience. A good first step to cracking such ciphers is to make a what is called a letter-frequency chart; after all, "e," "t," and "a" account for almost a third of letters used in written English. Knowing this gives you an idea which ciphered letters are really "e," "t," or "a," as well as others. You also have other clues; for example, a one-letter word will be "I" or "a"—assuming the spacing is maintained. The most common two-letter words are "of," "to," "in," and "it"; and the most common three-letter words are "the," "and," "for," and "are."

Some ciphers don't require "encipherment" as such, but are hidden within reams of unimportant text. In these ciphers, only one letter out of ten or twenty is meaningful. Finding the ones with meaning is like panning for gold: you sift through a lot of dirt to find your nugget.

The Engima machine produced highly complex ciphers for the encryption of German communications.

A stencil cipher is one example of this form. The message sender writes a long and misleading text, then the receiver covers it with a sheet that has holes in it. This covers the unimportant text, but leaves the important letters exposed.

Cracking the Enigma

Perhaps the most famous examples of ciphers were those generated by the German Enigma machine during the Second World War. The machine used rotors to generate complex ciphers for the encryption and decryption of secret messages. However, Allied decryption efforts at Bletchley Park, England—aided by captured Enigma machines—were successful, and are credited with hastening the war's end.

rotors

lampboard

keyboard

plugboard

Caesar and Vigenère Ciphers

These two ciphers are simple enough: they require only that letters are substituted with others in the alphabet. The Caesar cipher involves shifting each letter of a text a fixed number of places along the alphabet. The Vigenère cipher adds another layer of complexity by employing a keyword to unlock encryption.

The Caesar Shift

Named for Julius Caesar, the Caesar cipher is an extremely simple, and insecure, form of encryption. It is said that Caesar used a shift of three letters to encode his personal communication, so we'll take that as an example here.

Applying a shift of three to a text results in every letter moving three places along the alphabet; so an A becomes a D, a B becomes an E, and so on. In this way, the message "The party will be on November ninth" would be encoded as "Wkh sduwb zloo eh rq qryhpehu qlqwk."

Although effective for his personal communications, Caesar's cipher is not at all secure.

Increasing the Complexity

The Vigenère cipher takes this a step further, employing a keyword to determine the extent of the shift. In this example, we'll use three-letter abbreviations of the months of the years as possible keywords: Jan, Feb, Mar, etc.

To begin the encryption, a "tabula recta" is needed. This is simply the alphabet written out 26 times, each time beginning with a different letter. As you can see on the table opposite, the first line, listed as "A" in the table, begins with the letter A. The second line begins with B and is listed as "B," and so on.

A	B	C	D	E	F	G	H	I	J	K	L	M	N	O	P	Q	R	S	T	U	V	W	X	Y	Z
D	E	F	G	H	I	J	K	L	M	N	O	P	Q	R	S	T	U	V	W	X	Y	Z	A	B	C

	A	B	C	D	E	F	G	H	I	J	K	L	M	N	O	P	Q	R	S	T	U	V	W	X	Y	Z
A	A	B	C	D	E	F	G	H	I	J	K	L	M	N	O	P	Q	R	S	T	U	V	W	X	Y	Z
B	B	C	D	E	F	G	H	I	J	K	L	M	N	O	P	Q	R	S	T	U	V	W	X	Y	Z	A
C	C	D	E	F	G	H	I	J	K	L	M	N	O	P	Q	R	S	T	U	V	W	X	Y	Z	A	B
D	D	E	F	G	H	I	J	K	L	M	N	O	P	Q	R	S	T	U	V	W	X	Y	Z	A	B	C
E	E	F	G	H	I	J	K	L	M	N	O	P	Q	R	S	T	U	V	W	X	Y	Z	A	B	C	D
F	F	G	H	I	J	K	L	M	N	O	P	Q	R	S	T	U	V	W	X	Y	Z	A	B	C	D	E
G	G	H	I	J	K	L	M	N	O	P	Q	R	S	T	U	V	W	X	Y	Z	A	B	C	D	E	F
H	H	I	J	K	L	M	N	O	P	Q	R	S	T	U	V	W	X	Y	Z	A	B	C	D	E	F	G
I	I	J	K	L	M	N	O	P	Q	R	S	T	U	V	W	X	Y	Z	A	B	C	D	E	F	G	H
J	J	K	L	M	N	O	P	Q	R	S	T	U	V	W	X	Y	Z	A	B	C	D	E	F	G	H	I
K	K	L	M	N	O	P	Q	R	S	T	U	V	W	X	Y	Z	A	B	C	D	E	F	G	H	I	J
L	L	M	N	O	P	Q	R	S	T	U	V	W	X	Y	Z	A	B	C	D	E	F	G	H	I	J	K
M	M	N	O	P	Q	R	S	T	U	V	W	X	Y	Z	A	B	C	D	E	F	G	H	I	J	K	L
N	N	O	P	Q	R	S	T	U	V	W	X	Y	Z	A	B	C	D	E	F	G	H	I	J	K	L	M
O	O	P	Q	R	S	T	U	V	W	X	Y	Z	A	B	C	D	E	F	G	H	I	J	K	L	M	N
P	P	Q	R	S	T	U	V	W	X	Y	Z	A	B	C	D	E	F	G	H	I	J	K	L	M	N	O
Q	Q	R	S	T	U	V	W	X	Y	Z	A	B	C	D	E	F	G	H	I	J	K	L	M	N	O	P
R	R	S	T	U	V	W	X	Y	Z	A	B	C	D	E	F	G	H	I	J	K	L	M	N	O	P	Q
S	S	T	U	V	W	X	Y	Z	A	B	C	D	E	F	G	H	I	J	K	L	M	N	O	P	Q	R
T	T	U	V	W	X	Y	Z	A	B	C	D	E	F	G	H	I	J	K	L	M	N	O	P	Q	R	S
U	U	V	W	X	Y	Z	A	B	C	D	E	F	G	H	I	J	K	L	M	N	O	P	Q	R	S	T
V	V	W	X	Y	Z	A	B	C	D	E	F	G	H	I	J	K	L	M	N	O	P	Q	R	S	T	U
W	W	X	Y	Z	A	B	C	D	E	F	G	H	I	J	K	L	M	N	O	P	Q	R	S	T	U	V
X	X	Y	Z	A	B	C	D	E	F	G	H	I	J	K	L	M	N	O	P	Q	R	S	T	U	V	W
Y	Y	Z	A	B	C	D	E	F	G	H	I	J	K	L	M	N	O	P	Q	R	S	T	U	V	W	X
Z	Z	A	B	C	D	E	F	G	H	I	J	K	L	M	N	O	P	Q	R	S	T	U	V	W	X	Y

	A	B	C	D	E	F	G	H	I	J	K	L	M	N	O	P	Q	R	S	T	U	V	W	X	Y	Z
S	S	T	U	V	W	X	Y	Z	A	B	C	D	E	F	G	H	I	J	K	L	M	N	O	P	Q	R
E	E	F	G	H	I	J	K	L	M	N	O	P	Q	R	S	T	U	V	W	X	Y	Z	A	B	C	D
P	P	Q	R	S	T	U	V	W	X	Y	Z	A	B	C	D	E	F	G	H	I	J	K	L	M	N	O

Assume that 'we use "Sep" as' our key. This directs us to rows "s," "e," and "p" of the table. If encrypting a text, we would use row "s" to encode the first letter, row "e" for the second letter, and "p" for the third. For the fourth letter, we revert back to "s," repeating the process thus.

To take a line of text as an example, "our cover is blown, we need to make new plans," becomes "gyg uskwv xk fagac, oi cwis ls bsot fil hppfw."

BLAISE DE VIGENÈRE

Blaise de Vigenère was a French diplomat who lived from 1523 to 1596. He was the developer of a cipher system, but rather oddly not the one which bears his name. The Vigenère cipher was actually described by Giovan Bellaso in 1553.

Index
·········

Abel, Niels 27
Abu Kamil 68
Abu Tahir 72
Ahmes 59
Anaximander 34
Apastamba 62
Archimedes 21, 28, 46–7
Archimedian solids 48–9
  square root of three 38–9
Ariston of Chios 50
Aryabhata 30, 63
Augustine of Hippo, St. 6

Babylonians 28, 63
Bachet, Claude 54–5
Banu Musa brothers 68, 70
Baudhayana 62
BEDMAS 22
Bellaso, Giovan 141
Bernoulli, Jacob 125
binomial theorem 122–3
binomials 26
BODMAS see BEDMAS
Bombelli, Rafael 54, 95, 101
Bouguer, Pierre 25
Boyer, Carl 71
brackets 22–3
Brahmagupta 11, 15, 27, 63, 64–6, 84
Bürgi, Joost 136

Caesar cipher 138–9, 140
calculus 15
Cardano, Girolamo 76, 92–3
ciphers 138–9, 140–1
Clementi, Pier Francesco 101
codes 138

coefficients 26
combinations 115, 123
complex arithmetic 96–9
  and quadratic equations 102–3
complex numbers 94, 95
composite numbers 16
conics 74–5
conjugates 100–1
cosecant 31
cosine 31
cotangent 31
counting numbers 14
cubic equations 73, 93

decibel scale 137
degenerate conics 75
Descartes, René 11, 76, 95, 106–7, 110
Diophantus of Alexandria 54–5, 61, 64, 71, 101
dodecahedrons 43

e 125
Egyptian mathematics 58–9, 63
Enigma machine 139
equalities 24
Eratosthenes 46, 50–1
  circumference of the Earth 51–3
  sieve of Eratosthenes 50, 52
Euclid 8, 27, 28, 44–5
Euler, Leonhard 21, 124
Eulerian cycles 124–7
exponential equations 134–5
exponents 26, 132–3
expressions 24

factorials 112–13, 123
Fermat, Pierre de 55
Ferrari, Lodovico 101

Ferro, Scipione 93
Fibonacci 69, 80–1
Fibonacci sequence 7, 85–7, 89, 90, 120
Fior, Antonio 93
Fourier, Joseph 8
fractal geometry 94, 119
fractions 14–15, 59
"fundamental theorem of algebra" 129

Galileo Galilei 7
Gauss, Carl Friedrich 95, 128–9
golden ratio 88–91
Golenishchev Papyrus 59
graphing
  lines 108–9
  parabolas 82–3

Harriot, Thomas 25
Harun Al-Raschid 68
heptagonal numbers 18–19
Heron of Alexandria 39
hexagonal numbers 18–19
hexahedrons 43
Hipparchus of Nicaea 28–9
Hippasus of Metapontum 38
House of Wisdom, Baghdad 68, 70
Hypsicles 54

i 95
icosahedrons 43
imaginary numbers 94–5
  see also complex arithmetic
inequalities 24–5
irrational numbers 15

j 95
Al-Jayyani 30
Jones, William 21

Al-Karaji 69
Katyayana 62
Al-Khwarizmi 10, 30, 68,
    70–1, 76, 84
Al-Kindi 68

Laplace, Pierre-Simon
    62–3
Leonardo da Vinci 90
logarithms 136–7

Madhava 21
magic squares 63
Mahavira 63
Malik-Shah 72
Al-Ma'mun 68
Mandelbrot, Benoit 119
Menelaus of Alexandria 29
Mersenne, Marin 110
Mersenne primes 111
Mnesarchus 34
monomials 26
Moscow Papyrus 59

Napier, John 125, 136
natural numbers 14
negative numbers 14, 95
notation 11
number sets 14–15
number systems
    Arabic 69
    Babylonian 28
    Egyptian 58–9
    Indian 62–3
    Indian–Arabic 8, 9, 67, 85

octahedrons 43
octaves, notes of 35
Omar Khayyam 72–3
order of operations 22–3

Pascal, Blaise 110–11
Pascal's triangle 116–21,
    122–3
PEDMAS see BEDMAS
Pell's equations 64
pentagonal numbers
    18–19
perfect numbers 18
permutations 114–15
pH scale 137
phi (φ) 88–91
Philopater 50
pi (π) 10, 15, 20–1
Plato 40–1
Platonic solids 41, 42–3
polynomials 26–7
prime numbers 16
    Mersenne primes 111
    sieve of Eratosthenes
    50, 52
Prthudakasvami 63
Ptolemy 29
Pythagoras 34–5
    and music 35
Pythagorean theorem
    35, 36–7
Pythagorean triples
    37, 81
    square root of two 35, 38
Pythagoreans 35, 40, 42

quadratic equations
Al-Khwarizmi's method 71
    completing the square
    60–1
    and complex numbers
    102–3
    graphing the parabola
    82–3
    quadratic formula 76–7

radicals 122–3
rational numbers 14–15
Recorde, Robert 25

rhetorical algebra 11
Rhind Papyrus 59
Richter scale 136–7
secant 31
Sierpinski's triangle 118
similar triangles 29
sine 31
Singh, Simon 7
slide rules 136
Socrates 40
square numbers 16–17,
    18–19, 81
square roots 38–9
Sridhara 63, 76
stencil ciphers 139
sulbasutras 62
symbolic algebra 11
symbols, standardization
    of 25
syncopated algebra 11

tangent 31
Tartaglia, Niccolo 92, 93
terms 26
tetrahedral numbers 18
tetrahedrons 43
Thales 34
Theaetetus 42
Theon of Alexandria 54
triangular numbers 17,
    18–19
trigonometry 28–31

Varahamihira 63, 116
variables 26
Vigenère, Blaise de 141
Vigenère cipher 139,
    140–1

Wallis, John 21
Weber, Wilhelm 129
Wiles, Andrew 7

zero 15, 65, 94–5

Glossary

· · · · · · · · · · · · ·

Terms and symbols are explained where they are introduced within the text; however, a few are noted here for the sake of clarity.

Base the number or variable that forms part of a power and is acted upon by an exponent; for example, in the term x^2 the x is the base.

Coefficient a constant value that multiplies a variable; for example, in the term $3x^2$ the 3 is the coefficient.

Equation an equation always includes an equals sign; for example, $3x - 5 = 13$. Note, $\approx$ denotes "approximately equal to."

Exponent the number or variable that forms part of a power and indicates the repeated multiplication of the base by itself; for example, in x^2 the 2 is the exponent, and is equivalent to $x \bullet x$, while x^3 is equivalent to $x \bullet x \bullet x$; note that x implies x^1.

Expression a collection of numbers and variables that has neither an equals nor an inequality sign; for example, $(3x - 4) + 5$.

Inequation an inequation has an inequality sign where the equals sign would normally be; for example, $3(x + 2) \le 2x + 5$. Inequality signs include: $\neq$ (not equal to); $<$ (less than); $>$ (greater than); $\ge$ (greater than or equal to); and $\le$ (less than or equal to).

Like terms terms with the same type and number of variables; for example,

$6x^2$ and $8x^2$ are like terms, but $6x^2$ and $8x$ are not because the exponents are different, and neither are $6y^2$ and $8x^2$ because the variables differ.

Multiplication the symbol $\bullet$ is used in place of "x" to avoid confusion with the variable x. Multiplication is also indicated by closed-up numerals or variables. For example, both $x \bullet y$ or xy can be read as "x multiplied by y."

Operation a mathematical action; for example, addition or subtraction.

Plus-Minus indicated by the symbol $\pm$, within the context of algebra this represents two equations within one formula, and implies two solutions. For example, the formula $(x + 3) = \pm 7$ leads to two correct solutions $x = -10$ and $x = 4$.

Polynomial a collection of terms that include whole-number exponents. A polynomial of one term is a monomial; one with two terms is a binomial; and one with three terms is a trinomial.

Power strictly speaking, the combination of base and exponent; however, in everyday language it is often used to refer solely to the exponent.

Term a number or variable, or the product of several numbers or variables, that is separated from another term by addition or subtraction.

Variable a symbol, often x or y, used to represent a variable value in a term.